LEADERSHIP REFINED BY FIRE

LEADERSHIP REFINED BY FIRE

A Firefighter's Guide to Develop Leadership Skills,
Motivate and Inspire Others,
and Deliver Exceptional Care for the Public

JOHN M. CUOMO

WILD BULL MEDIA

www.wildbullmedia.com
contact@wildbullmedia.com

P.O. Box 670
Jupiter, Florida 33468

ISBN: 978-1-957652-00-9 (print)
ISBN: 978-1-957652-01-6 (ebook)

Ordering Information:
Special discounts are available on quantity purchases by corporations, associations, and others. For details, visit www.FDleadership.com.

TABLE OF CONTENTS

INTRODUCTION

It was a beautiful South Florida day—sunny, crisp, and cool—that started off like any other. I arrived at the station, checked the fire engine and equipment, had a little snack for breakfast. The officer in charge felt sick and was heading home. However, before he could be relieved of duty, we received a call.

Suddenly, the station's alarm tones rang and we were off: dispatched to a call of a "foreign object on the beach." For Palm Beach, a city with over 12 miles of beachfront, this was not unusual: unique items were always floating up on the shore. When we arrived on the scene my officer told me to run ahead and find out what is going on and start handling it. I suddenly felt very nervous. What do I say? How do I address the people on the beach? I was still a new firefighter with very little experience. Despite my nerves, I felt I could handle things. The other firefighter and I walked up to a crowd that had gathered in a circle on the beach. I approached the scene and worked my way through the crowd to discover, at the center, what appeared to be explosive military weaponry. A missile had washed onto shore! (see image on p.187)

My stomach sank and my mind went blank, I was shaken. The other firefighter on scene was as new as I was and kept asking me, "What do we do... What do we do?" I had no idea. I tried to think, but nothing was coming to me except more questions like: How do I get control of the scene? Is this thing going to blow up? Who do I call? My immediate focus was everyone's safety.

I radioed my officer and tried to explain what I was seeing. I believe he was in disbelief but began to make his way to the scene. I radioed dispatch and requested police presence, then asked the police to help secure the scene by keeping people far away from the object. That was a start, but I was still lost on how to proceed. One of the police officers suggested using dispatch to contact the military. That was a great idea, and I suddenly felt a sense of direction. As I began to radio dispatch, my officer finally arrived on scene. What a tremendous relief! *Thank goodness,* I thought, though I could tell that he was not prepared to handle the situation much better than I did.

In the end, we were told the object was a dummy missile used by the Navy for target practice. Apparently, this target missile had been missed and floated to shore. Although the call was for a dummy missile, I left feeling like the only dummy on the beach that day. It was an eye-opening experience.

A BETTER KIND OF LEADER

From the earliest days of my firefighting career, I was interested in leadership. When I saw that people's lives were in our hands every day and that an officer's decisions could place his crew in grave danger or keep them safe, I recognized that leadership in the fire service could be a matter of life or death. That day on the beach made me realize that it didn't matter what position I held; people relied on me just because I was a firefighter at an emergency scene. I knew I had to work toward becoming a better leader immediately.

I began to read books and take courses on fire service leadership, but I found they primarily focused on two distinct areas. The first was teaching you how to be a good fire officer by learning to read smoke, understand building construction, and fight fires. These are all essential parts of a good leader's skill set, as we will discuss later in this book, but there is plenty more that is needed. It seemed to me to be just the tip of the iceberg.

The second area I observed books and courses focusing on was personality typing—using tests and information to determine what type of person and manager you are. Do you have a type A personality? Are you a micromanager? Are you overbearing or too lenient? While I understood how my personality might affect my interactions with others, I needed more specific information. I wanted to learn how to lead a crew, especially in instances that were life-or-death situations for them or for those we were called to help. I wanted to learn how to care for the needs of individual crew members as well as the crew as a whole. I wanted to learn how to lead a crew, especially in life-or-death situations for them and those who called for help. I wanted to learn how to care for the needs of individual crew members as well as the crew as a whole. I wanted to know how to care for patients, deal with difficult subordinates, set the proper example, stay in good physical shape, and conduct workable relationships with city officials. Most importantly, I wanted to learn to search for my own deficiencies —and when I found them, tackle them effectively.

I studied these fields for years to try and find the answers I was searching for. I started putting together bits and pieces of what I thought would work well for a leader in the fire service. I spent an enormous amount of time dwelling on this topic, especially after I began stepping up as an officer just three years into my career. Suddenly, this 24-year-old kid had to make snap decisions in deadly circumstances.

My studies became more intense. I branched out beyond the fire service as I learned and put into practice what I saw working for great leaders and thinkers throughout history. I read about presidents like George Washington, Abraham Lincoln, Theodore Roosevelt, and John F. Kennedy. I researched successful military leaders like General Patton, Alexander the Great, Sun Tzu, and Major Dick Winters of the 506 Airborne, as well as exceptional businesspeople like Jamie Dimon, Oprah Winfrey, Warren Buffet, Henry Ford, Elon Musk, and others.

I could learn something from leaders in all facets of life, I realized, as I continued to widen my studies. I delved into the lives of great sports figures like Jackie Robinson, Michael Jordan, and Vince Lombardi, great motivators like Tony Robbins and Les Brown, and great human beings like Dr. Martin Luther King, Jr., Robert Kennedy, and Mahatma Gandhi. Further, I studied the writings of Victor Frankl, Carl Jung, Friedrich Nietzsche, Marcus Aurelius, and others to help me learn how people act and react. And I continued to be inspired by stories and people in the Bible. In addition, I listened to countless hours of audio and watched more videos than I can explain.

Every one of the individuals I've listed—and so many more—accomplished great things and had many lessons to teach. Most importantly, I had many things to learn from them. And I had many things to learn from my own mistakes along the way. I titled this book "Refined by Fire" to reference lessons learned from the fire service, but also areas of personal growth from difficult situations I got myself into.

So, this is a different type of leadership book than the ones I read from the fire service. I hope it contains universal insights for leadership in any field. This book's goal is to identify the qualities that have made great leaders in many fields, and determine how to apply them to the role of a leader in the fire service. **This book's most important function is to act like a mirror, to help you examine yourself against these leadership qualities to see what areas you can focus on, develop, and improve.**

Use this book as a game plan, the initial document in your personal War Room, your leadership command center. It was designed not only to learn from, but also to use as a reference guide you cancome back to throughout your career.

Can you lead like the world's great leaders? I believe you can!

Some names in this book have been changed to protect the privacy of those involved.

PART 1

LET'S GET STARTED

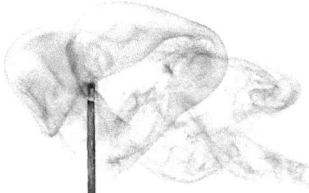

Chapter 1
THE JOB IS A JOURNEY

Are you reaching out to an office of oversight for the first time in your career? Are you already part of a management team looking to go to the next level? Or are you where you want to be but striving to be a better leader in that position?

The points we'll discuss in this book will help you accomplish all three, wherever you fit. But in getting started, understand that no matter what you are reaching for, the process of getting there will open up something new in your life. With something new comes change, a change in your responsibilities yes, but more importantly a change in you: what you do, how you do it, what viewpoints you hold, and how you express them.

It's been said, "If you want something you've never had, you must be willing to do something you've never done!" Seems logical, doesn't it? It is, and it's true. In the fire service, people will tell you it's not just a job, it's a career—but more than that, it's a journey. And as with any journey worth taking, there's a treasure waiting at the end. No, it's not your pension, although that will be there too.

The treasure is the ability to look back on a life that has been spent in the service of others.

It's the satisfaction you get when you've saved property from damage, you've saved the Earth from hazardous materials, and you've done the ultimate: you've saved lives. That is a treasure that cannot be measured, particularly for those who love the people you've saved. You can look back and know that your life has had a profoundly positive effect on those you've touched, both outside and inside the fire department.

Through the years, I have come to see that most firefighters who leave the service look back on their careers, what they accomplished, and the impact they had. You will too, I promise. Those who never truly understood that the fire service is more than just a job are *always* disappointed and unfulfilled. You will see these people too in your career: some longtime veterans who are bitter, angry, or always complaining. Don't let that be you.

There is almost nothing worse than regret. How do you prevent it from happening? Understand what you have entered into, what you are a part of. Understand that this career is a journey, with its ups and downs, peaks and valleys. There can be many downs, but if you keep moving toward your destination, you will get out of those valleys and back on a peak.

Are you ready for the journey? It starts with you. *But where? How?* You might ask. A famous Chinese saying states, "A journey of a thousand miles begins with a single step."[1] But I would suggest that any successful journey, even a vacation, requires four steps: Plan, Pack, Enjoy, and Reflect.

Have you ever gone on vacation without giving enough thought to planning and packing the appropriate things and regretted it on the road? Firefighting is a much longer and more difficult journey: don't start this journey out without planning.

Step 1: Plan and Prepare

How do you plan in this career? Think about the level you want to achieve, where you want to end up: firefighter, captain, chief? Ask questions of firefighters who are in those positions already. What do they feel is required? What helped them to achieve that position? What have they learned since being in that position that they didn't know before—how can that help you? Dig. What skills are needed? What training is required? What kind of commitment is it? Begin to self-examine, what are your strengths and weaknesses? Where do you need to give some attention to be able to be successful in that position you are interested in. What college education is needed? Some departments require a four-year degree for any management position.

What courses can you take that may not be required but may help you be better at your job? For example, you may wish to take an English course that will help you with report writing, or a speech course, which may help you in speaking to groups, residents, city officials or others.

Read books that might help you. Widen your range and diversify your scope of reading material. Read about great leaders in our history: many are quoted throughout this book. Read about their lives, what they accomplished, what made them say the things quoted here, what you can learn from them. Find material that will make you a better firefighter, a better driver/engineer, a better captain and/or a better paramedic.

Step 2: Pack

How do we pack? We pack for this journey when we study and apply the things we've learned in a positive way. Before a camping trip, for instance, you'll want to understand how to set up a tent, light a campfire with the tools you have, and what to pack for hiking in your chosen destination, so that as challenges come up you're ready for them.

Similarly, to "pack" and be ready for challenges as they arise in firefighting you need to learn from each experience you have and continually apply those lessons when you make everyday decisions. Make use of them so that they become tools that you have at the ready. They will help you avoid the valleys or, in any case, climb out of them to reach your destination. When you're suddenly hit with that missile on the beach or a difficult subordinate or a massive decision, the lessons you've learned and packed will help you navigate your way to a solution. You don't want to be the officer in charge whose mind goes blank and is visibly nervous and fumbling when you come upon a dangerous item washed up on the beach or a fire made more complex because of hazardous materials or another emergency. How many times will your crew look past your failure to be ready?

Step 3: Enjoy the Journey

There's an old saying: "Success is a journey, not a destination." Enjoy the journey! Don't let anyone or anything discourage you as you find your way—step by step. Know that you won't stay in the valleys if you don't want to be there. Acknowledge your mistakes and learn from them. Be proud of your accomplishments and build on them. Make note of the accomplishments along the way and keep moving forward.

If you do it right, you will accomplish great things, like improving human lives and having a lasting impact on those you work around and those you serve. Can you think of something you've already done that you can look back on with pride and satisfaction? What would it feel like to multiply that 100 or 1,000 times? Don't wait until the end of your career to look back at those moments; reach out each day to savor them along the way. They are the moments that will lift you up out of the valleys. Don't get stuck down there. Enjoy the journey.

Step 4: Reflection

When you return home after a great trip, you look at the photos and talk about your travels and reminisce about the joys and adventures. If you do the journey of your career correctly, you will look back on it with pride and contentment: you will reminisce about the great calls, the great friendships, the great achievements you witnessed—and the greatness of saving lives—for the rest of your blessed life.

Now let's work on that plan!

Chapter 2
PLAN. PREPARE.

Plan what? Prepare what?

Plan for your journey, plan for your career, plan to lead, to be a leader. Why? Because it is the best way to succeed.

As we go through life, we don't often stop and think about all the planning and preparation that's gone into the things we enjoy. For example, when we watch a favorite TV show, which lasts perhaps 45 minutes, we don't think about the hundreds of hours that went into planning and preparing it. When we read a book in a few hours or days, we don't consider the thousands of hours the author and a team put into planning, writing, and preparing it.

Examine how members of a football team prepare for success. Before they square off in a competition, they study every facet of the opposing team. They examine film footage of each player. They look for ways, even small ways, that they can use what they've learned to try something that might lead to a win. Then they practice all week for that game. That's a lot of preparation time for something that—albeit is a big business with a lot of money on the line—in the end, is a game. Win or lose the players all get to go home.

So can we possibly imagine that dealing with life threatening scenarios like fires, terrorist attacks, medical emergencies, mass casualty incidents, and other dire situations would not take significant planning to lead? What about completing a subordinate's evaluation, setting the right example, being a good subordinate yourself, staying physically fit, and hundreds of other tasks required throughout the day? Can those be completed successfully without planning and preparing?

A wise person once said that war is ninety percent information. What did they mean? They were speaking about information gathering, preparation, and planning. It was important to know everything they could about the enemy, the challenge, the numbers of troops, the weapons, where were they stationed, who the commanders were, what their styles of warfare were, even what great generals of past empires had done. In fact, Napoleon once wrote, "Read over and over again the campaigns of Alexander, Hannibal, Caesar, Gustavus, Turenne, Eugene, and Frederic... This is the only way to become a great general and master the secrets of the art of war."[1] Napoleon planned and he prepared, not only by learning everything he could about what he was facing, but also by using what he could learn from others before him.

How many leaders would you say are great leaders? History would name a relatively small number. Great leadership is a challenge, not just for others but for **yourself.** Marcus Aurelius, in his book *Meditations*, compared the battle you must wage within yourself to a boxing match.[2] Internal challenges can be some of the greatest challenges we face. They require—demand—a full commitment of our mind. This is a mantra I will repeat throughout this book.

The harder you work, the greater the results. And any accomplishment in becoming a better you—and a better leader of others—will result in positive outcomes throughout your life. It's not just your career journey that will be positively affected, but every facet of your life. Putting in the hard work is a win-win for you in every sense of the word.

WHAT CAN HELP YOU PLAN?

1. As you read this book, begin to implement the suggestions that work for you, and use the following steps to help guide you. Along with reading other firefighter leadership books, broaden your reading scope: You'll find a list of additional reading suggestions at the end of this book. Listen to audio books, watch videos, attend seminars on firefighting and leadership; do anything you can do to help in this path. Make a wholehearted pact with yourself that you will, from now until the end of your life, work on making yourself a better you.

2. Start the day right. To begin:
   a. Wake up being grateful.
   b. Empty out any negativity from the day before.
   c. Steer your thoughts and emotions in the direction you want them to go.
   d. *Carpe diem*—seize the day! Start Immediately.

3. Set a goal and keep your goal in focus.

4. Work on one goal at a time.

In all your life's endeavors—from the simple process of making a meal to the complex process of building a house—planning helps you succeed. You are about to embark on something very special. Use this opportunity to make a game plan to help you succeed.

Now let's lay the foundation!

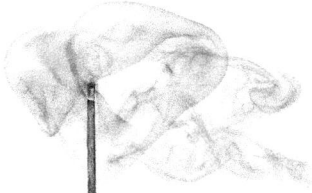

Chapter 3
YOU LAY THE FOUNDATIONS FOR YOUR SUCCESSES AND FAILURES EVERY DAY

"Yours is a me-first generation."

"Your generation is impatient and self-centered."

"It's a drive-through society."

"It's an I-want-it-now culture."

If I had a dollar for every time I heard one of those statements when I was growing up, I could have stayed home and lived a wealthy life. It's probably true that people today have a much shorter attention span than they did years ago. This is certainly reflected by the media we consume and our lifestyles. As I have gotten older and the internet and cell phone have come of age, these conveniences have made it harder for people to stay focused and easier for them to want instant gratification.

For example, we have actually moved from a drive-through society to a deliver-it-to-me society, and not just eventually, but today or in the next hour. It was not that long ago that the only way to contact someone was to call them via landline on a house phone. No answering machine, no call waiting. If the phone was busy you had to wait and call back and hope that the person would be home and off the phone. We had one phone in our home and it was on the kitchen wall. There were no cell phones, no email, no internet.

What does all this have to do with leadership? Well, it's more than just a nostalgic walk down memory lane of a simpler time. Everything in your life has an influence on you and your career as a firefighter: what you see, what you hear, what you read, the daily actions of others around you, everything. Thus, a society of people who are more impatient, who demand results much faster, and who have less ability to focus—especially long term—will have a definitive effect on us. To avoid picking up these traits, you have to fight against them. Why? Because what you are embarking on—becoming a great leader—will take time, and not just time over the next day or week but time throughout your career, coupled with consistency and focus.

Too much societal influence can cause you to give up on your pursuit too early, to quit during difficult times, to not give the full effort that's needed, to not put in the hard work every day over time to see results. When you do that, when you give up too early, it creates a weak foundation to build upon, which can lead to failure later on. How do you prevent being negatively influenced by things around you while working to achieve your goals? The title of this chapter is the answer. Understand that you lay the foundation for your successes and your failures *every day*. Let's discuss it.

First, what is a foundation and why is it important to building leadership skills in the fire service? The foundation is the base or beginning for all your successes, your goals, your achievements. But it can also be the reason for your failure.

In construction terms, the foundation is the most important part of a structure: a building's strength lies in its foundation. Its purpose is to hold the structure above it, keep it upright, and transfer its stresses to the ground. No matter how good the materials are that go into the rest of the structure, if the foundation is weak, the structure can collapse when strained. Laying a foundation is not a simple matter. The ground beneath it must be tested to guarantee stability relative to the size of the structure.

Most worthwhile things in life require a good foundation: marriage, raising children, your career, getting a promotion, succeeding as a good leader—the list goes on and on. Building that foundation takes time and effort. You've heard the saying that Rome wasn't built in a day. Rome grew to become an empire, one of the largest and longest-lasting in history. The point of the saying is not that it just took time to build Rome but that it took time and a tremendous amount of work, day after day, to build an empire! It takes consistent effort over a period of time to succeed, to create that good foundation. The more you work on the small things, the more disciplined you are and the more endurance you add to this, the better the likelihood that you will get what you want. You will be successful.

While this principle can and should be used for any and all major desires and ambitions, we'll focus it on the goal of becoming a great leader. There are many areas that need attention to achieve this, and we've already agreed that this process will take some time. So, understand that life is a series of events that build upon each other, one brick at a time. You cannot build a house in one day, but each day counts and counts a lot. Each day, the actions you take add up.

When I started my career as a firefighter, there was constant talk about a young lieutenant who was the protégé, the proverbial golden boy. He had talent and great ambition; however, he did not build the proper foundation. He grossly misunderstood the point of promotion. A promotion is just an opportunity to serve a greater number of people. The higher you are promoted the more people you serve. He apparently thought that with each promotion he attained, it meant a greater number of people had to serve him. He was promoted to driver, then lieutenant, then captain and then to assistant chief, despite much negative feedback from other firefighters and crew members who he mishandled during his career. Eventually, it caught up to him. His weak foundation cracked and he was asked to leave. He lost an incredible gift to serve others.

On the other hand, we had an officer who had a good reputation and tried very hard to be a good leader. One day, he made a big mistake in the way he handled a personnel issue. He was strongly disciplined for his error. He owned up to it, apologized and accepted his discipline without complaint. He seemed to work even harder to be an even better leader and the course of his career continued forward and upward. His foundation was tested and held because of his previous actions. And the way he handled this mistake actually ended up strengthening his foundation even more. He came out from this error a better and stronger leader and even more respected by the fighters and officers.

It takes time to build that strong foundation. Both of these officers worked on their foundations. Over time, the actions of the first officer made his foundation weak and, when stressed, it broke. The actions of the second officer made his foundation strong and, when stressed, it held.

No one goes from good leader to bad in one day. It happens in a series of events. A failure to set the right example, an action that tells the crew you don't really care about them, saying something negative about your boss, losing your cool and doing something stupid, showing up at an emergency scene and handling it badly because you were not well prepared—actions like these add up. Over time, your actions lay the foundation for your success or your failure.

We opened this chapter by talking about the dangers of societal influences—influences that could cause us to not walk the long road. It's human nature to stop doing something when we feel we are not seeing results or if results come too slowly. Making yourself into a true leader takes time. The good news is that once you start, you are on the road to success.

This is like a complex jigsaw puzzle: it takes time for the whole picture to come together. But keep working on putting the pieces together day after day after day and before you know it, it will become visible not only to you, but to those you lead and to others around you, including your boss, your family, and your friends. You will be a better person—guaranteed. And although these principles are designed to make you a better leader in the fire service, they will also make you a better mate, a better parent, a better son or daughter, a better employee, a better subordinate. It will influence you in all areas of your life.

So, let's sum it up. To be great, we must absolutely have a great foundation. We build that foundation with the right actions every day. What actions? Actions based on the qualities needed for leadership. We'll discuss these in the chapters ahead, but they include such strengths as love, courage, never lying or losing control, being balanced, and loving discipline, courage, and humility. Incorporating these qualities into our lives daily builds that great foundation. And how do we do that? We make them our habits.

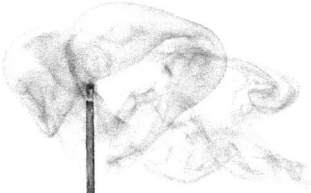

Chapter 4
HABITS: MAKE THEM YOUR LIFE, MAKE THEM WHO YOU ARE

"We first make our habits,
and then our habits make us."
— Unknown

Your habits are perhaps the single most important link to your success in any endeavor. Of course, we are focusing here on leadership in the fire service, but again, understanding this principle will help you succeed in all your life's endeavors. ***Your habits are the building blocks of your life.*** Let's break this down into a simpler step.

Football is a game that requires teamwork, just as our job does. One of the greatest football coaches to ever coach the game, Vince Lombardi, said, "Winning is a habit, unfortunately so is losing."[1] His catchy phrase helps us to understand that while habits are exceedingly important to our success, it's the *right* habits that we want to develop. The wrong habits will lead us down a destructive road. This is evident not only in the workplace but in life in general. Developing the actions and attitudes outlined in this book and making them your habits is an excellent starting point. And there may be additional habits that great leaders have used successfully that you will find in your own research. It's the right attitudes and actions that you employ each and every day at work and in life that will lay the foundation of your success.

So, what is a habit and what behaviors should we make our habits?

A habit is defined as "a manner of conducting oneself" or "an acquired mode of behavior that has become *nearly or completely involuntary*" or "a behavior pattern acquired by frequent repetition or physiologic exposure that shows itself in regularity or increased facility of performance."[2]

The key words here are "acquired behavior," "completely involuntary," and "increased facility of performance." As Aristotle once said, "We are what we repeatedly do. Excellence, then, is not an act but a habit."[3]

We've just discussed how what we do each day lays the foundation for our success. Now we can see how it pieces together. If you want excellence in your career as a firefighter, you want the good attributes, the good attitudes, the good actions—the things that great leaders do—to become so fully ingrained in you that they become a part of your automatic reaction. Without even having to think about them, they become who you are.

We can illustrate it this way: When you get into your car, do you consciously tell your hand to get the keys out of your pocket, then tell yourself to sit in the seat, then tell your hand to put the keys in the ignition and turn? Do you remind yourself to look behind you as you back up and tell your foot to slowly push on the pedal, being careful not to push the pedal all the way to the floor? Do you fight to keep your eyes on the road?

No, but think about the first time you drove, or even the first 30 times you drove. When you were first learning to drive, it was an intimidating task, but now it's automatic. You have repeated the steps so many times that they've become habitual, automatic. There are so many tasks that you accomplish all at the same time while driving, yet they are so ingrained in you, you actually do them without thinking.

How about the first time you drove the fire engine? Where you nervous?

Terrified? I was. Now you do it without even thinking about it. That is what is called developing habits: "making them your life, who you are."

THE HABIT OF LEADERSHIP

So, how do we make leadership traits a habit?

The first step is essential—you must whole-heartedly buy into the fact that attributes will make you a better leader.

Secondly, you must live the actions, day in and day out, down to the smallest behaviors, so that they become you. Politician, diplomat, and general, Colin Powell once said, "If you are going to achieve excellence in big things, you develop the habit in little matters. Excellence is not an exception, it is a prevailing attitude."[4] That is what we are trying to achieve, a habit of excellence! Being a great leader demands it! Leading people into situations where their life depends upon you and your decisions demands it! Representing the fire service demands it!

Aristotle, quoted earlier, explains it for us: if excellence is our goal, then it must be practiced habitually, every day. Do you have the will, the stamina, the fortitude to look introspectively, to see the areas that you need to work on, and to do the hard work until the traits you desire become habits throughout your career and life? The answer, of course, is "yes" you most certainly do. The simple fact that you are reading this book tells me and it tells you, that you can achieve excellence. Why? **Because you want it.** That is the first step.

My son once bought me the book entitled, *Benjamin Franklin's Book of Virtues.*[5] It's a small book but I found an excellent tool in it that helped me then and still helps me. I would like to share it with you. Mr. Franklin wanted to develop what he called important virtues to live by every day. He came up with a list for himself.

After self-examination, he realized how much he needed to work on these virtues. He devised a tool to help him. He wrote out the virtues in a book and next to each virtue he placed a day of the week. He would work on one virtue every day for a week. Each night he would examine his day's actions and any failures he would mark in the book under that day. His goal was to have no failures by the end of the day and then the next day start on the next virtue while continuing to work on the previous one.

I loved the idea, so I did some self-examination and picked some things I wanted to work on. I modified the idea so that I could work on a single issue until I was completely successful then move on to the next one. Here is what one of my charts looks like.

I have a book that I write in each night. I gave myself a chart per item. You can have several charts going at the same time, but I suggest working on one to begin with. Master it for at least two or three weeks before you add another one. Once you've mastered one it will be easier to maintain that one while you work on another. Mr. Franklin worked on virtues for himself. You can work on anything: breaking a bad habit or starting a great new habit to help you excel, or both.

The rest of this book focuses on things to examine ourselves against and to work on, adopting anything we feel is needed personally. Leading oneself toward excellence takes great effort: Leading others as well magnifies that effort. Allow this tool to help you excel in that leadership.

To help himself stay strong and continue to work on his virtues, despite failures, which are bound to happen, Mr. Franklin wrote encouraging quotes in his book that he would review to help him stay the course and fight on. I suggest you do the same. There will be times in your career when you will feel defeated or broken, it comes with a difficult job. Working on becoming a great leader will add additional pressure at times. Inspiring quotations can give you the motivation to pick yourself back up and continue the course.

Here are a few you may like, or you may wish to write in some that you have found to help revitalize you and keep you focused.

Vince Lombardi:

> "Winning is a habit. Watch your thoughts, they become your beliefs. Watch your beliefs, they become your words. Watch your words, they become your actions. Watch your actions, they become your habits. Watch your habits, they become your character."[6]

Here is a slightly different version that has been credited to various people, regardless it is a powerful thought for consideration to help you in your quest:

> "Watch your thoughts, they become your words; watch your words, they become your actions; watch your actions, they become your habits; watch your habits, they become your character; watch your character, for it becomes your destiny."

Stephen Covey, author of *The 7 Habits of Highly Effective People*:

> "Our character, basically, is a composite of our habits. Because they are consistent, often unconscious patterns, they constantly, daily, express our character."[7]

This one from President Theodore Roosevelt is excellent for encouraging you to fight on, staying in the arena of change:

> "It's not the critic who counts, not the man who points out how the strong man stumbles, or where the doer of the deeds could have done them better. The credit belongs to the man who is actually in the arena, whose face is marred by dust and sweat and blood, who strives valiantly, who errs, who comes short again and again, because there is no effort

without error and short coming, but who does always actually strive to do the deeds, who knows great enthusiasms, the great devotions, who spends himself in a worthy cause, who at the best knows in the end the triumph of high achievement, and who at the worst, if he fails, at least fails while daring greatly. So that his place should never be with those cold and timid souls who neither know victory or defeat."[8]

I love this one attributed to the ancient Greek philosopher Heraclitus in reference to battle. Are you that one that he speaks of?

"Out of every one hundred men, ten shouldn't even be there, eighty are just targets, nine are the real fighters, and we are lucky to have them, for they make the battle. Ah, but the one, one is a warrior, and he will bring the others back."[9]

Do not be overwhelmed. Accomplishing great things, huge things, can seem daunting, but they can be accomplished by breaking them down into small steps, small goals, small achievements, and building on each success.

Again, as we touched on earlier, "A journey of a thousand miles begins with a single step."[10] Imagine looking at the final goal 1,000 miles away and thinking, I have to walk all that way? It could seem daunting, impossible. But all you have to do is take the first step, just look at one step. Just accomplish one thing today. It's that is easy. Make that step, then focus on the next. In other words, focus on one thing at a time.

Please understand this critical point. Whether you like it or not, you will develop habits. Habits are fundamental to being human, they will happen automatically, sometimes without you even recognizing that a new habit has been formed. It's life. Some people develop negative habits like excessive TV watching, eating the wrong foods, developing a bad attitude, sleeping too much, drinking too much alcohol or too often, and so on. Work to make your habits ones that will work for you, that take you to the right places, that take you on that 1,000-mile journey if that is where you wish to go!

PART 2

IMPROVE YOURSELF

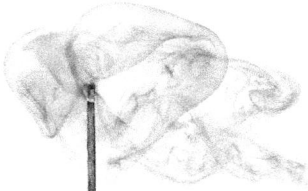

Chapter 5
KNOW YOUR JOB WELL

Know your job and know it well. This sounds simplistic but it's not. We will discuss four areas that will help you to understand how important this is for your ability to lead others and to see the need to never stop learning.

1 - Continue to assess

You need a significant amount of training before you can be hired by a fire department. In South Florida, you must go to EMT school, then fire academy and paramedic school before you have a chance of getting hired by a fire department. Some even get a fire science degree to help them stand out in the hiring process.

Once hired, though, many firefighters begin to dislike training: they stop training themselves and become annoyed or disgusted by programs assigned by the department. It's human nature. Think about all the things you did to win the one you love. Do you still do all those things to maintain that love: bring home flowers, cook favorite dinners, talk all night, stay in good shape?

Make an honest assessment of yourself right now in your career. Do you train and study as hard now as you did when you were trying to get hired? How well do you know *all* the aspects of fighting fire: engine equipment, medical SOPs, medical equipment, department rules and regulations? Are you in good physical shape to carry out your duties?

Next, begin working on the deficiencies you've found. Don't overlook the extreme importance of knowledge and training in your job. Don't risk allowing yourself, your crew, or your patient to get hurt because you didn't work hard enough at learning all you need to know. You might never forgive yourself.

Make these honest assessments continually in your career. Get it clear in your mind that you have accepted a career that requires constant training. Once you understand and accept that fact, it becomes easier to keep at it.

2 - Our job is ever-evolving

The COVID-19 pandemic taught us an important thing: the world is a dangerous place, and we don't always know what the danger will be or where it will come from next.

Back when I was in fire school, I remember the instructors talking about the enormous risks we face as firefighters. We spent three days discussing the great conflagrations in history. Conflagrations? Seriously? Since that time, American firefighters have had to respond to attacks on our country: planes hijacked and slammed into buildings on 9/11; bombings at the Atlanta Olympics, Oklahoma City, and the Boston Marathon; school shootings such as those at Columbine, Sandy Hook, and Parkland; mass shootings like the ones in Las Vegas and Orlando; church shootings; riots; ricin mail attacks; bird flu; swine flu; coronavirus; and so much more.

In addition, the Gulf and East Coasts of the country are threatened yearly with hurricanes including massive ones like Katrina and Sandy, and there are floods, hazardous material spills, car and truck accidents, water rescues, medical calls, and, oh yeah, fires.

Conflagrations? Please. I wish that was our big worry. If you stepped off a fire engine at one of the scenes I've just described, and you were in charge, could you run it? Could you lead? Our job is one of the most important in the country: keeping people safe, saving lives, helping people in trouble, protecting property—all the while risking your own life. Is it possible to be proficient at all of these things? Yes, it's possible. But to do so, you must know your job well and you must train consistently. Training is imperative to success at an emergency scene, but there is another critical factor to it as well.

3 - Knowledge is critical to leading others

Make the crew you lead your concern. They will know if you don't know what you're doing. They will be inspired with confidence by your knowledge or terrified and disgusted by your lack of it. The men and women you lead know that their lives and the lives of their patients are affected by your orders.

> *The single greatest factor determining your effectiveness in an emergency is how well you know your job. And without continual training, you cannot possibly know your job well.*

If you don't know it well, it can have disastrous consequences. I will never forget a department smoke-out rescue drill early in my career. I was new, and this was a training drill involving multiple fire departments. We were given a two-story house scheduled for demolition to work with. The windows were boarded up and the house was filled with smoke. Only one window was accessible. We were to climb a ladder to the second story, enter the house through the window, and begin a strategic search. It was hot, pitch black, and dense with smoke.

I hoped I would be paired with my friend, who, like me, was aggressive, competitive, and in love with being a firefighter. I loved being the best, having the best time, being first in training exercises, especially if it was a multidepartment drill. Unfortunately, I was not paired with my friend, I was paired with my lieutenant. He was an ornery man, and it didn't take much to set him off. Boy, do I have a lot of stories involving him.

He assigned me to be the wall man. Right from the start, it was evident to me that the lieutenant did not know how to do a search and that he was afraid. I could feel his apprehension and fear. Within the first room, I was already encouraging him and working to settle him down. I wanted us to clock the best time of the day, so I didn't want this drill to take too long.

The lieutenant kept complaining and whining. I kept shouting out encouraging instructions, telling him, "It's okay, we'll get through it, don't worry," on and on, trying to keep him cool, while at the same time—as a new firefighter—trying to do my job right. Finally, one of the instructors, who was following us using a heat monitor, asked him if he was going to be able to make it through. He said "yes" and quieted down, at least for a few minutes.

I could not believe what I was witnessing. We were firefighters. This wasn't even an emergency. It was hot, yes, but there was no fire. We cleared the second floor and made it down the stairs. The lieutenant was so incredibly slow, and I couldn't help but be frustrated. I kept thinking that our numbers were going to be horrible. If it had been a real emergency, how well would we have served the patient? What would have happened to me?

In the second room downstairs, I finally found the victim—a training dummy—and I yelled out. The lieutenant replied with a yell of his own, "Thank God, let's get the f#@k out of here now!" Instead of the two of us dragging the dummy together, I was now dragging it by myself and giving my lieutenant directions to get out at the same time. As we crossed the room, I found what felt like a reclining chair in my way. Instead of explaining what I encountered as I had before, I was now desperate to get that victim out and stop our time. Frustrated, I picked up the chair and threw it out of my way. (Early in my career, I admit, I was like a bull in a china closet. I loved to destroy. I'd rather go through a wall then through a doorway.)

I grazed my officer with the chair and that was the final straw for him. He freaked out. "Get me out of here, get me out of here," he yelled and began walking aimlessly. One of the trainers grabbed him and walked him out. I sat there in disbelief and angry. I pleaded with the other instructor to let me finish by myself, but he insisted I leave. I was embarrassed that we didn't even finish the drill. Forget the bad time, we had completely failed.

My friend, who was paired with another aggressive firefighter finished the drill and achieved the best time. I was upset, but what was worse was that I had completely lost all confidence in my officer. He did not know how to do a very basic part of his job. He did not apologize for his performance, explain what had happened, or work on his skills in the future. He was the rescue officer on a fire; this task we tried to perform was going to be his, and, unfortunately, my job on a real fire.

This is unacceptable, I thought. I told my wife what had happened and that I was afraid the lieutenant would perform the same way in a real fire. His lack of job knowledge would resurface throughout his career and continue to get him into trouble. No one wanted to work with him. When the department was finally presented with a good reason, they jumped at the chance to terminate him. Fortunately, no firefighters were placed in a position where his actions could have hurt them, but sadly some patients were.

His lack of job knowledge had a destructive side effect on the department as well. Besides being dangerous, it was a poison. It destroyed morale, it produced anger, it made people plot behind the scenes to get him fired, it affected patient care, and it created a spirit of separateness instead of teamwork. These effects are far more destructive than you might imagine, and they have long-term consequences.

Perhaps worst of all, this poison stopped all progress. The department lost an amazing opportunity to encourage its team to excel and be great, and instead set a terrible example indicating it would tolerate firefighters who don't know their job. The lieutenant regarded the absolute privilege of caring for patients with no value whatsoever, and therefore patients received terrible care at his hands. The incredible privilege to train men and women to excel was thrown away.

Do you know an officer who, from your observations, does not know how to do their job? Would you feel confident going into a fire under their orders or risking your life in another way? Don't be that person. Your crew will never respect you or your orders. Your department will suffer, losing an excellent opportunity to teach and create teamwork, and your patients will suffer from your lack of skills. It's impossible to be a great leader if you don't know your job well.

4 - Direct performance correlation

Your job knowledge and ability to perform will directly affect the lives you are called upon to help. When that alarm goes off in the firehouse, a person—or more than one person—has just been hit with something that is about to change their life. They are scared, nervous, anxious, and over-whelmed. They may be terrified and perhaps in excruciating pain. They are worried about the consequences of this event altering their lives in a drastically negative way.

Then you show up. Are you the one who brings the comfort, the help? Or are you going to make their life-altering event worse? Do you help mitigate the problem for them, or do you intensify it? Your job is to protect and save lives, conserve property, and serve the people of your city in the capacity *they* need it.

| *What level of service do you think people are expecting or hoping for?*

If you were diagnosed with a life-threatening disease, would you choose a doctor who just got out of school? Or would you choose a doctor who'd not only finished her schooling but also trained and specialized in this type of disease, including all the latest techniques to fight it? The choice is obvious. Don't we all want the best, especially when it comes to protecting our lives?

It's not hard to understand that the level of training dictates the success of doctors for the patients they serve. Why imagine it would be any different for firefighters and other first responders? Don't those we serve want the very best just as we would? Don't they deserve that from you and me? Isn't it easy to understand that our performance in emergencies is dictated by the level of our training and our job knowledge? If you can see that clearly, then the answer is simple. We must be proficient and continue to excel in our job knowledge and how to perform it. As a leader, you call the shots, you make the determinations regarding care or strategy that will dictate an outcome, both for your patients and your crew members.

So how do you achieve this level of excellence? By training, reading, and studying. Aristotle said, "The roots of education are bitter, but the fruit is sweet."[1] Malcom Gladwell wrote that it takes 10,000 hours to master something.[2] I am not debating the exact hours, but both men were right: Learning new things takes time, effort, and hard work—pain.

> *Training is the key to success. Training, reading, and studying. Every day.*

Let's look at the words of a few world masters of their trades—trades that are not only difficult but that require constant effort to stay on top. Anthony Joshua is the former heavyweight boxing champion of the world. The man truly looks like walking perfection. If I had to picture what Adam looked like in the Garden of Eden, it would be him. The following is a word-for-word excerpt of an interview with him after he lost to Andy Ruiz Jr. and was about to fight him again for a rematch, which Mr. Joshua would go on to win.

Interviewer: Will you be more nervous on fight night then you normally are?
AJ: No way
Interviewer: Because it's a different situation?
AJ: I am never nervous.
Interviewer: Never?
AJ: Look at the last fight. I was half asleep in the ring. (Speaking about the first Ruiz fight).
Interviewer: (laughs)

AJ: I am never nervous. I know I am good. When you know you are good at something you never make it the biggest thing in your life. *The biggest thing in my life is actually training.* Fighting isn't hard. The fighting and winning isn't hard, *the training every day, having the right people around you every day, that, honestly, that is the biggest fight in my life.* The winning is easy, that takes care of itself.[3]

He said that training every day is his biggest fight. He does it every day, he becomes a master at it, and then the challenge become easy.

Now look at this interview with the fastest man in history, Usain Bolt. He was being interviewed while he was training.

Usain: I think a lot of people, they just see you run and they say, "Oh, it looks so easy." It really looks effortless, but before it gets to that point, it's hard. It's like hard work, it's day in and day out sacrifice. It's day in and day out, just dying. There are times when you're running and you just want to stop. You just want to give up like, "To hell with this, I just want to go home." There's days when you get up and you know that you have a training today, you know it's going to be intense and you're like, "Oh, God, I don't want to go today." But you have to go. It's so hard and a lot of people don't know.
Reporter: We are not used to see[ing] you suffering.
Usain: That's why you guys are here, to show the true story, ain't it?
Reporter: This here is the true reality? Do you mean the competition is not the reality?
Usain: Listen, the work is behind the scene. The competition is the easy part. Behind the scenes is where the work is done, everything is done to get to that one race you need to run.[4]

The work is behind the scenes, before the challenge. In fact, the challenge ends up as the easy part.

Finally, notice what Mike Tyson wrote in *Undisputed Truth*:

> "The boxing's the easy part. When you get into the ring to fight, that's the vacation. But when you get in the gym, you have to do things over and over till you are sore, and deep in your mind you say, 'I don't want to do this anymore.' I push that out of my mind."[5]

Can we say that the emergency, whatever it may be, is easier than our training? If we do enough training, we probably could. Your training should always be done in order to learn something. Otherwise, don't waste your time or your crew's time. Train in full gear when it's appropriate. Read to learn something new. Study. Over the years you should assemble an enormous reading library. Atwood H. Townsend once wrote, "No matter how busy you may think you are; you must find time for reading or surrender yourself to self-chosen ignorance."[6]

> *You don't know what you don't know because you don't know that you don't know it!*

Watching videos and discussing tactics with others on your team is an excellent way of learning. You must put thoughts in your mind in order to translate them to your hands. Reading books on training can help too. Many colleges offer training classes, and most fire departments have training programs for you to follow. However, you can always go above and beyond and set up special training sessions for yourself and your crew and perhaps even your department.

I built many training props for my department and set up exercises that proved beneficial for all of us. For instance, firefighter survival was a field of study I was always keenly interested in, and one of the areas within that field is the RIT team (Rapid Intervention Team), also known as a RIC (Rapid Intervention Crew). The idea of this team is to be on standby, ready to respond at a moment's notice to a lost or down firefighter in an emergency scene.

Our department was small compared to those in larger cities, so extra personnel was a luxury we didn't have. For any big fire, we would call mutual aid to give us additional support. For years, I put in requests to allow me to buy the equipment and put together a RIT response policy. Finally, some years after becoming a lieutenant, the department gave me the green light.

I was to identify the pieces of equipment needed, order them, test them, set up the bags, practice with the equipment, write a procedure, and teach the department. I was thrilled. I immediately got my crew involved in choosing and testing equipment, and enlisted them further in helping to put the procedure together. I located an old furniture showroom in an adjacent city and got permission from their fire department training people to use it for our drills.

All summer long, my crew and I smoked out this large showroom full of smaller rooms and practiced every RIT scenario we could imagine. We became extremely proficient: our time, speed, efficiency, and knowledge were masterful. Next, we set out to train the whole department, and that was a phenomenal experience. I watched firefighters go from awkward to amazing, until what they were doing became as second nature and their confidence soared. Our crew became so close, it was awe inspiring. Their satisfaction with themselves, love for the job, and joy excelled. Other firefighters began asking to come train with me and work with my crew. The benefits we received from this can hardly be enumerated, and those benefits will spread far beyond us.

THE JOB NEVER STOPS CHANGING

As we've discussed, our job carries a heavy load. We are responsible for saving lives, and the job never stops changing. There is always something new to learn and master. Don't fret. The key is to take it in small steps. Master one thing at a time, then stay proficient at it by continuing to go back and refresh your knowledge.

Start today. Start practicing one thing today and tomorrow expand on it. Stay consistent and before you know it you will have become proficient at it, it will be second nature to you, or, as our three champions explained, it will be the easy part. It will become your habit. Get others involved, help them succeed and build teamwork together. Your confidence will bring you great satisfaction, it will attract the right people to you and it will help you succeed. Your skills will make you masterful at your job, whether it's keeping the crew safe, setting the example for them, or serving the public in the greatest way possible.

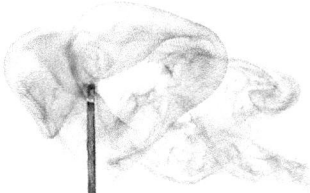

Chapter 6
PHYSICAL FITNESS

As firefighters, our job demands both intellect and physical fitness. We have to keep up with both. When people describe firefighters, they use words such as: hero, noble, strong, protector, brave, courageous, physically fit. No one ever says "weak."

Why does the public view us that way? It's not because all of us are physically fit or strong, or because they saw photos in a firefighter calendar. It's because *that is what they need us to be*. That is what makes them feel safe.

The public has no idea how difficult it is to drag and operate a hose with 100 psi of water flowing through it, or to be fully bunkered out and do a search and rescue, or to drag a victim out of a building, or to ventilate a home or commercial property. People don't realize what kind of shape you need to be in to swim in rough waters to rescue someone, or to carry on a hazmat operation in full hazmat gear, or to fight a wildfire. They don't know what is required of us physically, but they do know it is a lot.

When it comes to physical fitness, it's good to understand you have three obligations:

First obligation: your boss. When I say your boss, I am not talking about your direct superior or even the chief. I am speaking about the people you serve. I am talking about the taxpayers of your city or town or the county that you work for. Because our job is physically demanding, we owe it to our employers to not only know our job but to be physically fit. Would you pay your accountant if she could only do half of your taxes?

We sometimes forget that we were hired to do a specific, physically demanding job. Did you get in incredible shape solely to get hired by a fire department? Does it make sense after you were hired to let yourself go so that you are physically unable to do the job as well as you should? Should your employer, the taxpayer, pay you a salary to do a job that you can't fully do?

There is evidence that our physical condition affects our job performance on every call we're on, including medical calls.[1] Would you want someone who is unable to effectively perform CPR on your loved one to show up when you need help? Obviously not.

This doesn't mean you will always need be in the shape of a 23-year-old full-time athlete, but if you want to lead others in the fire service, you should be able to perform at a basic physical level.

Second obligation: your brothers and sisters. One of the greatest advantages and joys in the fire service is that we act as a team. As a team we are constantly relying upon each other, which draws us closer together, like family.

Most of the jobs we work on cannot be done alone. We physically need a partner. This gives us an obligation to our partner and to our crew to be able to pull our own weight. It's unfair and morally wrong to allow ourselves to get out of shape in this job. It will make it harder for us to do our job while putting our brothers and sisters and our patients in danger.

If you are the leader, that obligation multiplies and deepens. In that role you are responsible for more than just yourself. You are responsible for setting a good example for the team, a responsibility you accepted when you were promoted. By accepting that job, you said you would lead in all manners of your job. Don't you agree that a physical job requires leading in a physical manner as well?

That does not mean you have to be the strongest or the fastest. It means that you set the example, that you are in shape and are able to lead your crew, physically, in the job presented. You will certainly lose the respect of those you attempt to lead and others in the department if you can't. In addition, how do you evaluate your crew members on physical fitness if you are out of shape? How do you evaluate them in other areas and tell them they need to improve if you are physically out of shape to the point that you cannot perform your tasks in a physical job as well as you should?

Third obligation: yourself. You have accepted an offer to work in what is a physically demanding job. You know this. Those you lead know this. You must be true to yourself about what you do. If you cannot physically do the job, it will eat at you internally. No one likes to feel like they cannot do a job, especially when they see others around them able to do it.

There is nothing like the satisfaction of going out with a team, working until exhaustion accomplishing a task, and then coming back to the firehouse to clean up. After you clean up, there is a feeling of wholeness and satisfaction. You feel a closeness, a bond with your crew, you did a tough job and you did it well. Contrast that with being unable to complete what you were supposed to do, or having others pick up your slack. That is a terrible feeling.

We had a call in a very high-end hotel at a wedding reception. The patient, an older gentleman, was having chest pains. He was lying on the floor when we arrived, surrounded by people. We tried to evaluate him, but his drunk, extremely obnoxious son was all over us. He kept yelling at us, grabbing our equipment, grabbing his dad and shouting, "We don't go down, right dad? We are the Smiths, we don't go down!" All control of the scene was lost.

The lieutenant decided to load the patient onto the stretcher and get him into the back of the rescue vehicle so we could work on him in peace. We quickly loaded the man and headed out with his son draping all over the stretcher, still yelling about his family prowess. We came to a flight of stairs, about 15 straight down and plenty wide enough. Four of us grabbed the stretcher, one on each corner. Halfway down the stairs the lieutenant let go and the patient tilted toward that corner. He was strapped in, but it caused a major imbalance.

The three of us tried to stabilize the stretcher as the man flailed about, almost falling out of the stretcher. The two crew members not holding the stretcher, dropped the equipment they were carrying and rushed to our aid. We stabilized the stretcher and got it down the stairs. The patient's son exploded. The patient exploded. "Get me off this fu***** stretcher! Get me the hell off this fu*****stretcher now!"

The lieutenant apologized and tried to get him to allow us to continue to the rescue vehicle. He flatly refused and proceeded to verbally rip into us. The scene was chaotic. A bunch of people, concerned for the gentleman, had followed us out of the reception. We looked like a bunch of incompetent idiots. It was awful. Everyone was yelling at us and finally the gentleman got off the stretcher and walked back into the reception.

The terrible feelings we had about the scene were only intensified when we got back to the firehouse. Our lieutenant called a meeting immediately and then proceeded to yell at the crew and blamed everyone: the cop, the son, the patient, the people, the music, us—everyone but the true culprit, himself. Yes, it was chaotic, yes it was a difficult situation. But that is exactly what we do. We do difficult situations.

Ultimately, he had failed. He dropped his corner because he could not physically do it. We would have looked past it if it had just slipped out of his hand—mistakes do happen. But when he failed to apologize or even explain and instead blamed everything else, including us, the crew became irate.

After he walked out, the crew was furious with him. They wanted to send an email to the chief and get this lieutenant disciplined. Eventually everyone calmed down, but they lost a lot of respect for the lieutenant that night, and it was obvious he had lost a lot of respect for himself as well. He walked around like a humiliated kid for quite some time. It must have eaten at him for a long time. The guys joked and laughed about it afterward at his expense, even in front of him.

And what about the patient? Did we alleviate his harm or did we cause more? Obviously, we caused more. He didn't get the help he might have needed that night, and he lost trust in the first responders sent to help him.

Staying in shape benefits you and the way you feel every day. It also helps you stay around longer for those you love and who love you.

Every firefighter should be able to physically do their job, it's a necessity, especially if you are a leader. How can you possibly be looked to as a leader if you can't do the job? We are fortunate that our schedule, in most departments, allows us some time to work out during our workday. Some departments even have small gyms. There's no excuse to be physically unable to do the job when you have the time and can choose the gym instead of watching TV or playing on the internet.

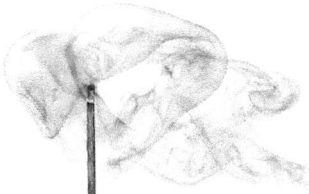

Chapter 7
NEVER LIE

Mark Twain once wrote, "If you tell the truth you don't have to remember anything."[1] My whole life I have heard people repeat something similar as a reason to never lie. It's a weak reason at best, and at worst it's a snare that covers up how much devastation lying can cause.

You will make many mistakes during the course of your career; it comes with being an imperfect person. I have made many mistakes myself. Some mistakes may haunt you for many years and some will be forgotten almost immediately. Some you will have to live down, meaning that you will have to demonstrate it was, in fact, truly a mistake, not who you really are.

Unfortunately, the bigger the mistake, the faster news of it will spread throughout the department, no doubt with a bit of additional creativity on the part of the spreaders. Time, hard work, and persistence will help bury mistakes, especially if you have already built up an excellent reputation in the department.

One mistake that may be impossible to recover from is lying and trying to cover it up. Why is it so hard to recover from a lie? *It's because a lie is not an accident, it's intentional.*

Perhaps the single most important thing I would want to impress on you would be to never lie. *Ever.* How can people follow someone they know is a liar? How can they ever trust anything that person says? A good leader can never be a liar and a liar can never be a good leader.

Of all the things we talk about in this book, this is the one that can destroy you and those around you. It is my personal belief that lying destroys you as a person, it rots your soul. It places you upon a slippery slope that leads to depths of ugliness and pain you don't want to feel. Does this sound too strong? We'll examine some evidence together, but first let me share this experience with you.

I was behind the fire station, working on building a training prop. It was a typical hot day in Florida, a Saturday, a day when the administration was off. The firefighter working with me that day was not one of my normal crew members. He approached me and said that he had hurt his hand during his last shift (three days earlier) and didn't report it. Now it was swollen and it hurt really badly. He asked if I would mind saying that I saw him hurt it today.

I felt a knot instantly develop in my stomach. This was not going to go well and I hoped he was just joking.

"I can't do that, that would be a lie," I said.

He looked at me stunned and began angrily protesting. "How could you not back me up, what's wrong with you? What kind of brother are you?"

He invoked the "brother card" to try and get me to lie. However, in reality, if he was my brother, he would have never asked me to risk my position and my career to lie for him.

So, I asked him emphatically, "You want me to lie for you?"

"Yes," he said.

I have learned that when you reframe a problem for people, they will usually see the wrongness in their actions and correct it. Well, that didn't work this time. Lying must not be a big deal to him. So, I calmly tried to reframe it again and asked, "So you are asking me to put my career, my family's livelihood, and my reputation on the line because *you* hurt your hand and *you* didn't report it, as *you* were supposed to do?"

I still did not reach him. He became frustrated with me and turned to walk away.

I said, "File the paperwork and if you hurt it last shift then say that and I will sign it."

"The department will deny the claim," he said.

"Well, we don't know that for sure, but if they do, you still have health insurance and paid sick time if needed, and more importantly, you didn't lie," I replied.

He gave me a look of disgust and walked away.

Thirty minutes later, he came back outside and told me I had a call from my boss, the battalion chief. (This was not uncommon, especially on weekend mornings the BC would call to see if there were any needs.) My boss proceeded to tell me to fill out the paperwork and write down that this firefighter hurt his hand this morning during the morning check. You see, the firefighter was good friends with my boss, so he had decided to go over my head.

I asked my boss, "Do you know he is lying, that he stated he hurt it last shift?"

He said, "Yes, but if we say that they will deny the claim."

I explained again that we don't know they will deny it but if they do deny it, the firefighter still had health insurance and a sick day route, if needed.

The battalion chief answered, "Just do what I said."

"I am sorry, I absolutely will not lie, you can write me up," I said.

Of course, that caused problems between me and my boss, unfortunately, for the rest of his career.

The firefighter wrote it up the right way and the department covered it. I called the firefighter into my office for a meeting that night. I told him not to ever ask me to lie for him again or to go over my head. I then explained to him the importance of never lying, the commitment to moral excellence, the pride of being a firefighter, the trust issues, and the fact that he could lose his job for lying.

I could tell that I didn't seem to get to him. Since we didn't normally work together, I didn't see him much, but when we did see each other, he was cold toward me. About two years later, he called me on the phone, frantic. I asked him what was wrong. He told me he was under investigation for lying and that he thought he would lose his job. He'd lied about something he was specifically told not to do and was caught. I told him to be truthful, to apologize, and to promise he would never do it again. I don't know exactly what he ended up saying, however he was immediately terminated. Apparently, he had been caught lying several times and this was the last time they would put up with it. What I found intriguing in this situation was that, out of all the people in the firehouse to ask for help, he called me. We were only acquaintances, we never worked together as a crew, and he was nothing but cold toward me.

I ran into him several years later at a social function. His wife had left him and he was in pretty sad shape. I asked him if she left him because he was fired.

"Well, that added to it," he said.

Obviously, I don't know all the reasons, but I do know this, the way you conduct yourself in the firehouse is the way you will conduct yourself in your outside life and vice versa. Working to better yourself has benefits not only in your career but your whole life.

My boss, the battalion chief in this scenario, would also later face more career problems for a variety of things, including lying. Does this mean that if you lie you are destined to be fired? No, what it means is that if you lie you are apt to lie again, and again. In fact, it has been shown scientifically that it becomes easier to lie the more you do it,[1] and it also leads to larger transgressions.[2]

Lying will destroy your character, period. It will rot you from the inside out and it leads to bad things. Is this true? Well, you be the judge, let's look at what your body says, what it tells you about lying.

THE POLYGRAPH MACHINE (AKA LIE DETECTOR TESTS)

Polygraph machines work by monitoring multiple physiological activities in your body in the following manner. Two expandable tubes are placed around your chest to monitor your breathing. Then two stainless steel plates are Velcroed to two of your fingers to create a circuit. And finally, you are hooked up to a blood-pressure cuff.[2]

When you tell a lie, your body reacts, and it is these reactions within the body that the detector picks up. Lying tends to make people sweat, meaning that sodium ions emerge from the pores, causing electrical activity in the circuit. Our fingertips are among the more porous parts of our body, perfect for locating telltale beads of sweat. In addition, when you lie, your blood pressure goes up and your breathing rate increases. This is an instantaneous response by your body. Interestingly, pain has a similar physiological response.

What is your body telling you? That lying is destructive to you personally. It's wrong. It is showing you that on the inside. But on the outside, your body language broadcasts when you are lying as well. Your body betrays you when you betray it.

Joe Navarro, who worked for the FBI and studied and wrote extensively about body language, states, "Those who are lying or are guilty and must carry the knowledge of their lies and/or crimes with them find it difficult to achieve comfort, and their tension and distress may be readily observed. Attempting to disguise their guilt or deception places a very distressing cognitive load on them as they struggle to fabricate answers to what would otherwise be simple questions."[3] He then goes on to expound on all the things the body does from this distress and discomfort.

These things tell you that it is against human nature to lie. You are going against or damaging your body and mind. Like drinking poison. Would you drink a glass with one tiny drop of poison, not enough to kill you, but to do a tiny bit of damage? How about doing that several times a day, every day. What is the eventual result? You destroy who you are. This analogy can be applied to lying. It may not destroy you physically, but it destroys your character, it rots your soul, and it will destroy all your relationships.

Jordan Peterson, in his book *12 Rules for Life*, dedicates a chapter to this topic himself. It is well researched and I believe comes to the same conclusions. He took it a step further to show how lies are so insidious that it was in fact lies from ordinary people that allowed for the destruction inflicted on the world by the Nazis and Communism in the 20th century.

While discussing this he points out: "Viktor Frankl, the psychiatrist and Nazi concentration camp survivor who wrote the classic *Man's Search for Meaning*, drew a similar social-psychological conclusion: deceitful, inauthentic individual existence is the precursor to social totalitarianism. Sigmund Freud, for his part, analogously believed that "repression" contributed in a non-trial manner to the development of mental illness (and the difference between repression of truth and a lie is a matter of degree, not kind). Alfred Alder knew it was lies that bred sicknesses. C. G. Jung knew that moral problems plagued his patients, and that such problems were caused by untruth. All these thinkers, all centrally concerned with pathology—both individual and cultural—came to the same conclusion: lies warp the structure of Being. Untruth corrupts the soul and the state alike, and one form of corruption feeds the other."[4]

Can you afford to deny all these great thinkers? Can you afford to deny the physical evidence your body gives you when you lie? Lying corrupts you, and it corrupts the fire department.

The fire department is a job of constantly changing factors. No two calls are exactly the same. No two calls can be handled the same. Different situations require different actions, even from minute to minute during an emergency, and they demand constant and continual judgment calls. This leaves a door open for mistakes to happen, or even perceived mistakes. Sometimes your decisions are made on assumptions. Sometimes you're just making a best guess.

A lie, though, is neither an assumption nor a guess. A lie is intentional. It is devious. It is covering over the truth about something. Lies destroys trust. Those you interact with will question what else you are covering up? What else are you lying about? Your motives will be questioned in all the decisions you make. You will be telling others by example that it's okay to lie. In the earlier example of the firefighter who was fired for lying, did the battalion chief aid in his eventual firing by telling him it was okay to lie? I believe so. The BC missed an excellent opportunity to teach the FF not to lie.

If someone should ask you to lie for them and you do, you have not only hurt yourself, but you have hurt them as well in many ways. When you have a reputation that you won't lie, people won't ask you to do so. In addition, they will have deep respect for you and your leadership.

What if there is a possibility that the consequences may be too severe if you don't lie? Even if the consequences are the direst—as in, you may be terminated—I personally believe the right thing to do is never lie. If you make a mistake, just fess up to it, accept any discipline, and learn from it. You may be surprised at how incredibly proud of yourself you are in such a situation.

For example, imagine you made a mistake that may cost you your career. If you face it— terrified of the consequences for you and your family—and yet you still admit it, you will build in you an incredible strength of character. It is something that can never be taken away from you and it will make you grow. In addition, you will earn an immeasurable amount of respect from those you lead and those who lead you. They will remember it. Your example will forever change them.

And what if you did lie and are sorry that you did. Then what? Admit it. Don't make an excuse for it. Apologize if you hurt anyone in any way. Ask for their forgiveness. Go to your superiors and explain it, if necessary, and promise it will never happen again. Sit your crew and your subordinates down and explain yourself to them. And when you have made amends to the others, make a vow to yourself that you will never go down that road again and walk away from it. Do not beat yourself up over it to the point of not moving forward and don't let anyone else beat you up over it.

Bottom line: never lie. It goes against our nature. We know deep inside it's wrong to lie and when we do it, we destroy a little part of us, our humanity. We didn't think enough of others to be truthful. *We didn't think enough of ourselves to be truthful.* We become a coward in our own eyes. This lack of humanity, lack of moral excellence, will show up in our other behavior as well if we let it. If you lie once, even if you get away with it, there is a good chance you will lie again, and then again. Hold yourself to excellence. Remember it takes years to build a good reputation and minutes to ruin one. Be proud to represent the fire service honestly.

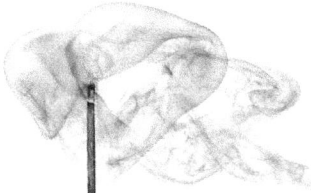

Chapter 8
NEVER LOSE CONTROL

Ironically, just before I sat down to write this section, I watched a video of a bunch of Pee Wee Football coaches who got into a fight after a game, right in front of the children they were trying to teach about sportsmanship, dealing with adversity, working as a team, and love of the game. Imagine one of those coaches was you. How do you live that down?

The video included clips of other coaches from all kinds of games getting into fights with each other. I am not pointing a finger at them to judge them. As I have said, I have made mistakes in the past myself. But I have tried my hardest to learn from my mistakes and this video was an excellent way to learn from others' mistakes —a lot less painful than learning from your own!

I have to believe that the men in these videos are not inherently violent people, but for some reason they lost control of themselves that day. The aftermath was certainly devastating. In some cases, aftermaths can be life altering.

Working in the fire service comes with immense daily pressure. Let's start with working 24 hours straight, sometimes 48 hours or longer. Lack of sleep. Years of accretive internal anguish from constantly seeing people

LEADERSHIP REFINED BY FIRE

in pain and suffering and death. Fire department politics, rumors, back-stabbing, and personality conflicts. What seems like never-ending nego-tiations with city offices over reduced benefits. Constant new procedures, rules and regs, memos, protocols, tests, training exercises, and tools. Then add life: problems at home, financial pressures, raising children, family health issues or deaths, working a second job, and your own imperfections and internal struggles.

You can become a powder keg ready to explode with the next spark.

All the ingredients are there for you to lose control, and the sparks are provided daily. Let's discuss the danger and what we can do to help prevent it as much as we can.

DANGERS

It has been said that you will not be punished for your anger; you will be punished *by* your anger. How does anger, or your loss of control punish you? The Roman emperor and philosopher Marcus Aurelius described it this way: "Anger surrenders an individual's autonomy to the offense or judgment of others. Anger represents an 'involuntary spasm' and a momentary lapse of reason. It severs the rational bond that unites all human beings."[1]

Anger makes us lose reason and causes involuntary actions and words.

I was hired by my department right after my 21st birthday. I was a kid, somewhat newly transplanted from New York, working as a firefighter in a Southern city. It felt like many of the guys held some sort of prejudice against me for being "from the North." I was like water to their oil; we did not mesh. To make things worse, I was stupid, arrogant, cocky, and I had a chip on my shoulder.

Instead of keeping my mouth shut, as I should have, I responded strongly to their harassment and didn't control my attitude. Those mis-takes would be held against me for some time. It didn't take me long to figure out what was happening and that the best course of action was to not respond at all. There were many times when I wished I could start my career over in a new department with a clean slate and a quiet demeanor. I

56

did end up applying elsewhere and was offered a position, but ultimately, for several reasons, I decided to stay. It would take more than half a decade for those early stories to finally die and reset how people viewed me. Sometimes it's just the way you respond that can hurt you.

At the time that I applied for my job at the fire department, the field was very competitive. It was routine to show up for a test hiring process to a department with 3,500 to 4,000 applicants when they were hiring for a single position. It took several years of working full time, going to school for the fire department, taking test after test for every city I could, before I finally got picked up. When it finally happened, it would be an understatement to say I was ecstatic.

I have always been someone who likes to show my gratitude for people who do something for me. I felt grateful that this chief hired me, but I did not have any idea what to do to show that gratitude. I asked my sister, and she said buy the chief a bottle of wine. So, before my start date I bought a bottle of wine and went to deliver it. He was not there so I gave it to the assistant chief and asked him to give it to the chief as a thank you gift from me.

I started one week later. By that time, the story had spread to the whole department. However, the story that spread was that I bought the chief a bottle of wine, not as a thank you, but instead to kiss up. In addition, the story had already grown. It was now a case of wine and a cake. Within the next two years the story would grow to a case of extremely expensive liquor, a cake, roses for his wife and dinner for them both. This was what happened when I did something to express my appreciation. Imagine what would have happened if I had lost control of myself and did something really bad.

When you make a mistake, and you will, because none of us are perfect, news of your mistake may spread throughout the department and grow. This can affect promotional opportunities and even your ability to lead others.

Another danger is when we lose control of our words. The Bible (NIV) calls the tongue "a fire," and speaks of what little fire it takes to set a whole forest ablaze.[2] When we lose control of ourselves, we have a tendency to say things we shouldn't and it can cause significant destruction all around us. It can destroy trust and the bonds we've developed.

As a leader you are aware of the faults, weaknesses, and vulnerabilities of those you lead. If you were to lose control of yourself and say hurtful and painful words about those things, you would destroy any bond you have with your team. They will never trust you for anything. In addition, embarrassing someone or making them look bad in front of others does not build up but tears down, does not exude leadership but exemplifies a petulant child who lacks self-control.

Once spoken, words cannot be taken back. Individuals will remember what you said. It can and will create a chasm between you and them. In addition, your words can spread to others, perhaps with the same vulnerabilities. These people too will see you in a negative light and certainly not as a leader. You will have an uphill battle trying to lead them if that moment comes.

When you lose control, you show your own weakness. As a leader in the fire service, you are expected to lead and manage people through some very critical times. It's not easy running fully bunkered out into a dangerous fire. It takes mountains of courage, and if those you lead see you as weak, you make it much more difficult for them to mount the courage to follow you into the flames. It's impossible to lead effectively like that. It takes work to maintain control of oneself, especially in the job we have. Sooner or later, it's likely you will fail. We all do. Some of us must fight harder than others not to fail. I've wished so many times I had the temperament of my children. But I do not, so how do we work on never losing control? Let's examine a few steps.

WORK TO CHANGE WHAT COMES OUT OF YOUR MOUTH

"Do not let any unwholesome talk come out of your mouths, but only what is helpful for building others up according to their needs, that it may benefit those who listen," states the Bible (NIV).[3] This verse in Ephesians lays out two steps for working on what comes out of your mouth. First, make it a habit to talk about good things, not bad. Sometimes we just plain slip up, but it's less likely to happen if we've made a habit of not speaking disparagingly about others. So, make it a habit to speak well about them.

The Bible (GNT) has more to say in Luke, "The mouth speaks what the heart is full of."[4] What we feel in our heart will come out through our mouth, especially when we lose control. We cannot feed into those bad things. Like the physical heart that gets filled with blood and then pumps it out, the metaphorical heart gets filled with what we put in there and pumps it out. Do we put hatred, jealousy, rivalry, racism in our hearts? Or do we fill it with love for others, peace, compassion, empathy, and other wholesome qualities? Put good things in your heart and guard it. If you do, you will have better control over the destruction your words can cause.

EXAMINE THE ROOT OF THE FEELINGS AND FIX THEM

When I was writing this chapter, I got a little snappish with my family. I try very hard to constantly work on myself, and had already noticed this, but when my wife gently pointed it out to me, it stung like a hive of bees had hit me all at once. I never want anyone, but especially those I love so much, to see any negative qualities in me.

I immediately went to my office and meditated on what was going on. I pinpointed the issue. I was having some difficulties with my extended family, but I, unfortunately, was taking it out on my immediate family. I set a game plan to work on it. Just recognizing what was irritating me inside and changing how I let it affect me was enough to change my whole disposition. It was no longer gnawing at me.

We want to continually evaluate our actions. When we say or do something, even a small thing, that is not the way we usually act, it's a warning sign that something is wrong and that if we don't find it and fix it, we'll lose control on a much larger scale eventually.

EXERCISE

Mental exercise works the same as physical exercise: the more you do it, the stronger you get. Maintaining control over yourself takes effort. When you begin to feel anger, or you feel yourself getting upset, police yourself, remember that you can maintain control. With each success, the stronger you will get. Be conscientious of what you are feeling. Tell yourself before you enter a situation that you must maintain control—it will help you. The more you do it, the more it becomes automatic, even if you sometimes fail.

NEVER MAKE EXCUSES OR JUSTIFY YOUR ACTIONS; OTHERWISE, IT IS EASY TO DO IT AGAIN

When you make excuses or justify your actions, you're telling yourself that you don't need to change your behavior. Without change, the behavior continues to get worse, and one day you'll lose control, leading to a full-fledged conflagration.

When you lose control, admit it, especially to yourself, and then work on not letting it happen again. When we lose control and cause pain for ourselves and others, it's usually the culmination of many smaller similar incidents that slowly grow until they swallow us up. Taking accountability for each smaller loss of control and working to fix it will help ensure that the big ones never happen.

YOU CREATE YOUR OWN EMOTIONS

No one can make you upset or angry or happy or glad. Those are all emotions that *you* yourself create from some sort of stimuli. Someone can do something to you that isn't nice, but it's you who chooses to create the emotion of being upset or mad.

You don't have to get angry. I know this sounds easier than it is, however, it can be done. There are ways to reframe things or rethink them in order to control your emotions. The better you get at this, the easier it is to do when things come up suddenly and there is no time to think.

It's extremely important as a leader to have control of yourself. People don't want control exercised over them by someone who can't control themself. They don't want to be led into life threatening places by someone who can't control themself. And working with someone who is always running at the mouth is exhausting, defeating, depressing, and nerve racking.

Self-control is essential. You will never perfect it, but you can get very good at it. Take responsibility for your actions. Fix it, pick yourself up, and keep working not to fail again. Work on changing what comes out of your mouth and what is put into your heart. Examine what makes you feel certain ways and fix the things that aren't working. Exercise mental control. Make yourself strong at it. Never justify when you mess up. It's never anyone else's fault but yours. Remember it's you who creates your response to the stimuli of other people or things.

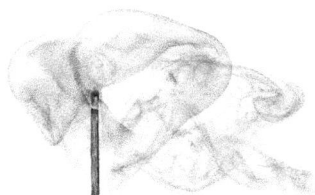

Chapter 9
ASK/GIVE: THE IMPORTANCE OF FEEDBACK

The word "feedback" is frequently used, but its power and impact are rarely appreciated. It's an extremely powerful tool in your leadership toolbox. Think of it like giving a flashlight to someone walking through a forest in the dark of night. With that flashlight they can stay on the worn path and avoid pitfalls or getting lost in the dense brush. It will help them avoid cuts and bruises along the way and help them achieve their ultimate goal, whatever it may be.

The better the feedback you give or receive, the brighter the flashlight. Hopefully, this book will help enable you to provide that as a leader, pointing out the stumbling blocks to those following you, offering encouragement, and helping them to stay on the right path.

Curiously, feedback is one of the most misunderstood and underused tools we have. I've read that it's not praise, advice, or evaluation—that it's only used to help someone reach a goal. But that misses so many of its facets, sort of like using your flashlight in the woods only to look up and failing to see the road ahead. Let's look at three important aspects of feedback.

Delivery - Delivery encompasses both verbal expressions and non-verbal actions. Feedback has an almost symbiotic relationship with its delivery. For example, you may hear it given as positive feedback or negative feedback, but some may say those are linked to its impact. For example, you may have heard a statement such as, "His constant negative feedback brings the group down." Those you deliver feedback to can and will have a reaction to it, and your delivery style will add or detract from the way it's received.

Impact - No matter what the feedback is or how you deliver it, there will always be an impact. With a good impact, the person on the receiving end may be motivated to action or change. With a bad impact, a problem could get worse, a bad attitude could develop, or perhaps some respect for you is lost. Of course, in answering simple questions, your response is just information, but those are not the types of interactions we are discussing here.

Actions - *You are constantly giving feedback, even if you don't mean to or realize you're doing it.* For example, your lack of action on something, a roll of the eyes when someone is speaking, jumping to your feet to address an issue, the way you keep your uniform and your gear, how you speak about others in the department or talk about your superiors. Every one of these actions, and so many more, send feedback to those around you and those you intend to lead.

Understand that you send messages all day, every day. Feedback is ubiquitous, let's see how you can use it for good. First, give as much feedback as you can. Remember that good feedback is like giving your subordinates that much-needed flashlight. Help them to be better. Help them to excel. Inspire them to aspire. Here are some examples of how you can do that.

When you have a new subordinate assigned to you, or if you are assigned to a new crew:

1. *Sit with each one individually.* Make an appointment to sit with them, one-on-one, privately. Make it as personal as you can,

maintaining professionalism. If you go into the meeting with that mindset, you will be more effective. It will let them know that you care about them but also that work is a business that has to be run professionally.

2. ***Let them know what you expect from them.*** It is difficult and frustrating for subordinates when they don't know what is required of them. Yes, there are standard operating procures—rules, regulations, and protocols—that they need to know, but your team should never be left guessing how you want things done. In the fire service, you will continually work with different officers throughout your career. Each officer has their personal way of doing things. Don't make a subordinate figure out what you want done by trial and error—weak officers will keep their subordinates in the dark for a variety of reasons. When someone knows what is expected of them, it is much easier for them to accomplish it and progress. This is good for the fire service and the customers.

3. ***Set up goals.*** Ask them what their goals are. Where do they wish to see themselves in three years, five years, 10 years, or even at the end of their career? What are their one-month goals, one-year goals, and five-year goals for reaching that destination? Then see how you can assist them in achieving their goals. This will help them make progress, become an integral part of the department, and make the department better. It will give them the confidence to keep progressing and make them want to follow your leadership. In fact, they will more than likely use your example to help their subordinates when they are promoted. Thus, you've begun to help set up a culture of leadership with every subordinate. Your example will have an effect on the department as a whole and the fire service itself.

4. ***Ask if they have any questions***, and if they do, answer honestly.

5. ***If anything is left open-ended***, make another appointment to sit down together to finish the meeting. Schedule it right then to

ensure that it happens. This also shows you're sincere in your desire to help.

6. ***Continue to give feedback throughout the year both formally and informally.*** Understand that the greatest feedback you give is through your daily actions, whether intentional or not. If you truly believe the things you speak about and try to live them, your actions will speak volumes. A picture speaks a thousand words, but your actions speak a million.

EVALUATIONS

Many officers are fully or partially responsible for their subordinates' yearly evaluations, whether it involves their raises or not. If you are in charge of evaluations, here are some suggestions:

1. ***Meet more than just once a year.*** Make specified dates to meet with your subordinate and go over their progress, work records, goals, etc. Having a quarterly review might work well, or more if you wish. No firefighter wants to receive an evaluation and find bad marks on it. When they sit down with you to go over it and ask, "why didn't you tell me about this earlier," what will your answer be? It's not my job? If you think that, you are sorely mistaken. It most certainly is your job to tell them.

   Consider the subordinate's point of view. If they desire to make advancement, you will be viewed as a hindrance to their success if you don't speak up. Certainly, they'll believe you don't care about their success.

   Please note: People are not stupid. If you are not honest, if you have a hidden agenda, or if you treat people differently because of their race, gender, sexual orientation, or religion, they will see through the games that are being played with them and that ultimately will affect their pocket and their advancement. Besides hurting a member of the fire department and ultimately failing at your job, you could be putting yourself in an extremely dangerous position. An employee's evaluation is not something to use to get back at a subordinate or spitefully hurt them because you don't like them for some reason.

2. ***If you are not responsible for evaluations in your department, celebrate.*** That lifts an enormous amount of difficulty off you. However, you can still try to help a subordinate succeed. Perhaps you can speak to the person who is responsible for the evaluations and find out what they are looking for so you can help your subordinate meet those goals. This is appropriate even for subordinates you only work with from time to time due to the fire service scheduling.

   The firefighter will never forget that you helped them. It will make them want to give back by doing a good job. More importantly, it sets an example for them to follow later when they are in charge, thus improving the fire department or fire service as a whole.

   Whenever possible, I gave deserving subordinates who worked with me a letter of thanks or accommodation to use on their evaluation. People appreciate being appreciated. It makes them want to work harder for you and makes them like their job more. One of the greatest responsibilities you have as a leader is to help others love their work. If they do, they will do a great job.

Well, we've been talking about giving feedback, now let's talk about something very important to *your* self-improvement, advancement, and happiness in the department: *Asking* for feedback. All the same principles apply.

1. Ask to meet with your supervisor shortly after you are assigned to him or her. If you will be having a yearly evaluation, ask to meet at the three-month, six-month, and nine-month marks to see about

progress and ensure you are both on the same page. These meetings do not have to be long, in fact you do not want to take more of your supervisor's time then necessary.

2. Ask the supervisor what they expect from you and respectfully express what your expectations are. This will not only eliminate frustrations from misaligned expectations, it can help establish a good working relationship right away. The goal isn't to get more money on your evaluation—although that will be a byproduct—it is to excel at what you do in the eyes of your supervisor. The best way to do that is to know exactly what they are looking for—and, of course, to perform all your regular firefighting and leadership duties well.

3. Goals. Be prepared with your short-term and one-year goals and be ready to talk about them. In addition, I recommend thinking about your three-year, five-year, and 10-year goals. You do not have to share those with you supervisor, but you may wish to. You might also ask if they have any goals for you to work toward and ask them to help you achieve those goals.

A good leader will relish the opportunity to help you achieve your goals. Giving them the opportunity to do so will cement a good working relationship and create an openness that helps prevent any surprise problems.

But what if your supervisor is not a good leader? What if they could not care less about your goals and really do not want to help you? That is bound to happen many times in your career. You must still try to work with them. And at the very least, you will know what it's like for your subordinates when an officer doesn't give a damn about them.

In any case, I promise that your efforts will have a positive effect on your supervisor, even if you don't see it right away. They most certainly will institute positive changes, if not when you're in the room, then perhaps with their next subordinate. I have witnessed it. Everyone wants to do a good job, or at least be thought of as doing a good job. Your leadership in the fire service can be an influence in so many ways.

4. Be open to scheduling another meeting before your next evaluation occurs. The more opportunities you have to meet and discuss things with your officer, the easier it will be to get on the same page and move forward.

So, give as much feedback as you can. Be generous with it. Just as we need food in order to grow physically—we need feedback to continually grow in our jobs. Take advantage of feedback yourself, whether or not it's given voluntarily, and even when it's not given kindly. Sometimes you may need to disregard the messenger and listen to the message. Try to maintain a communicative atmosphere with your subordinate and your supervisor. The more open and friendly it is, the better the feedback will be.

As we said at the start of this chapter, getting feedback is like being given a flashlight for a journey in the dark. If you give your subordinates that flashlight and ask your supervisor to give you one too, you will both reach your desired destination.

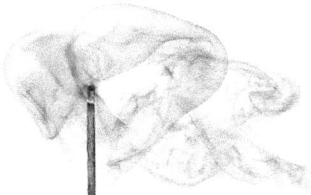

Chapter 10
FINDING THE BALANCE

Good things are accomplished in life by way of balance. Take food, for instance. It's essential to life, but too little of it and you become sick from malnutrition, too much and it leads to hosts of problems that rob your life. Balance is perfect. Water is also essential to life but too little of it and you become dehydrated, which can lead to death. Too much of it and you throw off your electrolytes, which can kill you. Balance leads to progress in exercise, temperature, sleep —you name it. The key to life is balance.

There is only one thing, in my opinion, that you can do excessively with no detriments and that is to truly love others. Love as much as you can, as many people as you can, and to the fullest degree that you can. Make the world a better place.

In the fire service, balance is essential. There are not too many occupations like ours where things can go from mundane to chaotic in seconds. Your response to the chaos can mean lives. Balance affects your approach to an emergency and your mitigation of it. Handling an emergency in a balanced way will help you see clearer, think faster, and be more productive. Those you lead will see it too. It will inspire them with confidence and pride in you and set an example for them to follow.

You can see that the world functions at its best when it is balanced, and so do we. Work through this journey in a balanced way. Not too little, not too much. The journey requires that you work hard and consistently but be careful not to let work dominate all other things in your life. Just as you feed and hydrate yourself every day, work at improving yourself—making yourself better—and a better leader—every day, even if it is just a little.

Set aside time to read something new, learn something new, or work to correct something about yourself every day. Having a specific time will help, and day by day your efforts will add up. Before you know it, you will have accomplished that goal, read those books, learned something new, fixed that deficit, and you won't be burnt out or exhausted from the journey. Then you can begin to work on the next thing.

In my childhood, I had learning disabilities. I could not read well, so I never read at all, but I recognized I had to get better at it. I found a book I thought I would like and began reading it. It took me forever to finish it. Then I picked up another one and another one. It was amazing to see my grades get better and better. I got to the point where I really began to enjoy reading: It's amazing how it opens up worlds to you and how much it helps your brain. Now if I don't get time to read every day, I feel very disappointed, like I missed out on something really good. There are so many things you can learn from such a variety of books.

And here is where balance comes in. Read this book and others and do a thorough self-examination. Determine how you want to improve and take it a bite at a time, don't attempt to do it in a day or a month. If you attempt to do too much at once you will fail, you will be exhausted. If that leads you to quit, you will be back where you started, or even worse off than before.

Work hard, but don't forget to relax and rest and spend time with those you love and who love you. We can get too focused sometimes, neglecting others around us. Working too hard and burning out, or worse, can cause difficulties in our relationships.

Being balanced will teach you patience. It will teach you that time is essential to the long haul, the big picture. *Balance will help you achieve your goals: others will see the differences in you, and you will as well. It will help you accomplish great things without losing yourself or the ones you love.*

So, decide today what you are going to work on. Write it down and come up with a balanced way to accomplish this goal. The core of this book is self-examination, personal responsibility, it starts with you.

I like these two photos; they show several things. The first is that you can balance many things at once, but not *too* many things. The second is that when you have many things to balance, just one small movement the wrong way can cause everything to collapse. That is where habits, patience, endurance, and hard work come in. They act as the glue that holds everything together, not only preventing you from knocking things over, but allowing you to balance more. Start small.

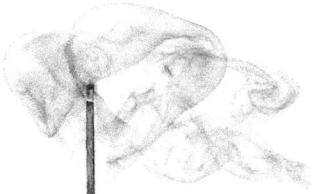

Chapter 11
EMBRACING FAILURE

"You may encounter many defeats, but you must not be defeated. Please remember that your difficulties do not define you. They simply strengthen your ability to overcome."
— Maya Angelou[1]

As we stated at the start of this book, firefighting leadership is a journey, and you will absolutely have failures along this journey. Just as on a long hike, you will stumble, slip, get sunburned, and take the wrong path from time to time. In terms of your job, that translates as getting written up, saying stupid things, making big mistakes and wrong decisions, failing at a goal, and feeling foolish sometimes. It will happen. You may even be treated unfairly at times.

You cannot make the journey without the failures. But just as on that hike with many setbacks, there's true joy and satisfaction at the end. The skinned knees and sunburn and near failures didn't stop you, they propelled you.

FAILURE IS THE KEY TO SUCCESS.

You may ask how can that be? Let's look at three ways this is true.

1 - Every failure if handled right, brings you closer to your ultimate success.

Michael Jordan once reflected, "I have missed more than 9,000 shots in my career, I have lost almost 300 games, 26 times I was trusted to take the game winning shot and missed, I have failed over and over and over again in my life, *and that is why I succeed.*"[2]

Michael Jordan is one of the greatest athletes of all time. What propelled him was far more than his God-given talent, it was his competitive spirit, his drive; he rigorously did the things that would eventually lead to his successes. However, as amazing as those successes were, they came with many failures. The key was how he handled them. He didn't let them stop him. He wasn't embarrassed or ashamed, nor did he let his failures make him feel small. He learned from them and kept pushing forward.

2 - Failure makes you stronger if you get back up and continue forward.

Booker T. Washington stated, "Success is to be measured not so much by the position that one has reached in life as by the obstacles which he has overcome."[3] Failures in our life are obstacles that sometimes stop us from moving forward. When we get back up, overcome the obstacles, and move forward in spite of them, we succeed. And when we look back and see what we did, it makes us stronger to overcome the next obstacles.

I wrote earlier about how it took me years and dozens of fire department tests to get hired. It was agonizing and debilitating at times. When I was finally hired by a department, it truly felt like the greatest success of my life. The obstacles, for that 19- to 20-year-old kid, had seemed impossible to overcome, yet despite so many setbacks, I accomplished it. It was a lesson that would propel me in the future to never give up.

Fredrich Nietzsche wrote, "That which does not kill us makes us stronger."[4] But this familiar statement isn't true unless you add, "as long as you get up and try again." That is the only way it works. Scientists at Northwestern University's Kellogg School of Management studied the effects of failure in the early years of an individual's career. They found that "failure early in one's career leads to greater success in the long term *for those who try again.*"[5]

> *Fail early, fail often, but get back up and keep trying and you will succeed beyond your dreams.*

3 - It makes you smarter, better equipped to continue on and face new challenges.

Henry Ford wrote, "One who fears the future, who fears failure, limits his activities. Failure is only the opportunity more intelligently to begin again. There is no disgrace in honest failure; there is disgrace in fearing to fail. What is past is useful only as it suggests ways and means for progress."[6] Embrace your failures. Learn from them. Failures have led to the greatest successes in our lives. Other people's failures and the way they have responded to those failures have significantly altered your life for the better.

Thomas Edison's teacher told him he was "too stupid to learn anything."[7] He went on to hold 1,000 patents for some of the most life-changing products. Isaac Newton's mother pulled him out of school to make him a successful farmer, but Newton failed as a farmer and was placed back in school. He would become a mathematician, physicist, astronomer, and theologian—one of the most influential scientists of all time. Vera Wang failed to make the 1968 U.S. Olympic figure-skating team. She became an editor at Vogue but was passed over for the editor-in-chief position.[8] She began designing wedding gowns at age 40 and today is one of the premier designers in the fashion industry, with a business worth over $1 billion. Walt Disney was fired from the Kansas City *Star* because his editor felt he "lacked imagination and had no good ideas."⁹ He would go on to build a media conglomerate based on imagination. Oprah Winfrey was publicly fired from her first television job as an anchor in Baltimore for getting "too emotionally invested in her stories."¹⁰ She used that formula to create a multibillion-dollar media empire.

There are so many more examples.

You will be judged not by your failures, but by the way you respond to them.

Take those failures and learn from them. Have you ever done something wrong and it was made public, to your embarrassment? I'm sure you never made that mistake again—perhaps you even told yourself that at that time. That illustrates the power of learning from failure. Learn not only from your own failures, but from those of others.

Failure can also help you learn to work harder. The first fire department job interview I had was at Delray Beach, Florida. I was 18 and had no idea what I was doing. After the interview, they told me they really liked me but because I didn't have my certifications, they couldn't hire me. They told me to get my school certifications and come back, and then they would hire me. I did just that. I went to EMT school and Fire Academy and got my certifications.

Shortly after, Delray Beach was hiring again. After working so hard to get through the schools and wanting so badly to get hired, I was super excited. Here was my chance, they had specifically told me to come back with my certifications and they would hire me. I applied and made it to the interview process again. But, instead of being hired, I received a letter in the mail stating that I was not being considered for hiring at this time, but I was on a list and they would call me. That call would never come. I was crushed.

What happened? Why didn't they hire me? Did I do a bad job on the interview? I had no idea, but I wasn't going to let it happen again if that was the problem. I was determined to figure out how to do a great interview. That failure pressed me to work harder on my interview skills and that would serve me well for the rest of my life.

> *Getting back up and trying again after a failure helps you to conquer fear. Fear becomes less of an obstacle in your life, all around.*

Throughout your career, regardless of your position, you are going to fail and make mistakes. You may feel foolish at times. More than a couple of rules were added to the rules and regulations book because of failures of mine: do not park your vehicle anywhere near the unit bays, do not take the truck out of zone for training, and several more.

It's okay. It happens to everyone. Don't try to lie your way out of it when you're caught in a failure. Don't try to pretend it's not what it is. People see through that. Own it and move forward. Learn from it and don't repeat it. People will respect that, and you will grow from it, you will learn from it. It's another rung on the ladder of success for you.

View any failure you may have as an opportunity not only to learn what not to do but also to become a better you. Steve Jobs would say his getting fired, from Apple, the company he founded, was a bitter pill to swallow but it was the best thing that could have happened to him.[11] He was able to focus on what went wrong and fix it. As history shows us, he came back and built a phenomenal company, a trillion-dollar company with products that revolutionized the cell phone industry, so much so that everyone copies them. In essence, his failure and handling it right helped all of us. The same holds true for many others who failed and used the failure to propel themselves to new heights. You can do it too. It is all in your hands.

"Even when I was close to defeat,
I rose to my feet."
— Dr. Dre, "Still D.R.E."

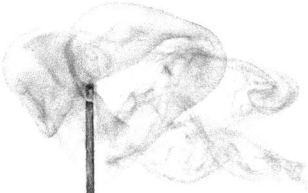

Chapter 12
THE ROLE OF MENTOR

The importance of having a mentor in the fire service cannot be overstated. A mentor can stop you from going down the wrong road, help you achieve leadership skills, save you from terrible pitfalls, and help you reach your career goals in and even outside the fire service. That doesn't mean you can't succeed if you don't have a mentor, but it certainly helps.

Imagine you're an 18-year-old about to enter college. Your desire is to work in the technology field, computers particularly. If you've been at this crossroad or have a child at this point, you'll know that finding the exact right direction initially is like playing the lottery once and picking the winning numbers. Yes, it happens, but the chances are one in a million to get it right.

The sheer ubiquity of computers in our society dictates an overwhelming number of directions that one could go in for study. What direction will the initial study lead to? Will the technology you learn be outdated by the time you graduate? What is the new technology coming down the road? Who are the best instructors and what are the best schools in your field? What are employers looking for in the different fields? Which computer language is best for your field? Which hardware or hardware developer is the right one to use? The list goes on.

But imagine you bumped into Bill Gates and he said, "I will help you, come work alongside me for the next couple of years and I'll guide you in your education and progress." Mr. Gates could give you invaluable insights and answers to all of your questions. Can you imagine the pitfalls you could avoid, the time you would save, the knowledge you could attain, and the speed at which you could get to an elevated level?

Now change the scenario. Imagine at 18 you wanted to learn how to invest in stocks and Warren Buffett offered to mentor you, or you desired to get into banking and Jamie Dimon was your mentor, or you wanted to become a tennis pro and Serena Williams offered to help you, or you wanted to become a body builder and Arnold Schwarzenegger was there to mentor you. Can you see the benefits of having a mentor? All these greats had mentors, too.

Arnold Schwarzenegger's mentors: Joe Wieder, Jim Lorimer, and Sargent Shiver[1]

Jackie Robinson's mentors: Carl Anderson, an auto mechanic, and Reverend Karl Downs. Robinson stated, "I supposed I would have become a full-fledged juvenile delinquent if it had not been for the influence of two men who shared my mother's thinking."[2]

Jamie Dimon's mentor: "Jamie Dimon saw Sandy Weill for what he was: a minor league legend on Wall Street who was offering the young man a ride in the express elevator to the executive suite."[3] This is the exact scenario I spelled out earlier and it was a tremendous help for Mr. Dimon's career.

Serena Williams's mentor: Billie Jean King[4]

John Maxwell's mentors: he has been teaching leadership for 50 years. He listed seventeen mentors he has had throughout his life, all in different areas.[5]

It may be difficult in a fire department, especially a smaller one, to find a mentor. But you may be able to find guidance in several areas from different individuals. You may notice that someone is an exceptional strategist, another may be exceptional at getting others to perform at top level, another may be in incredible physical shape, another may be excellent with customer service. Glean something from each of those individuals, looking to their strong suits and learning from them. Further, if you train with other municipalities and come to know excellent firefighters in different departments, you may find mentorship there.

STUDYING THE GREATS

You can also be mentored by others through studying their lives. The lives of John F. Kennedy, Robert Kennedy, and Martin Luther King Jr. have had a profound and significant effect upon my life. All of them died before I was born, but reading about them has changed my life. Their love for humanity and efforts to ease the suffering of others has been forged in my head and heart. I tried as much as I could to bring those attributes to the fire service when I served others. Even now, although I am not serving at this time, they are top priorities of mine.   Although nothing is as effective as having an actual person to talk to for guidance, I feel I've been mentored and have learned much from the many individuals I've read about, especially when they've told their stories in their own words.

Mentors can help you with more than just your career. They can teach you some of the most important life lessons: how to deal with tragedies, difficulties, and unfair treatment; how to treat others, care for them, and show love. Mentors can help you find what is in your heart and exemplify that. Reflect on just a few examples:

In 1948, at about 19 years of age, Martin Luther King, Jr. entered Crozer Theological Seminary to continue his education. He was confused as to how to attack the social issues faced by the country at that time. He wanted to use what was in his heart to help other people. Do you know where he got his "non-violent" approach from?

He said that he began "a serious intellectual quest for a method to eliminate social evil."[6] And he turned to a serious study of "great philosophers from Plato and Aristotle down to Rousseau, Hobbes, Bentham, Mill. All of these masters stimulated my thinking—such as it was," King said, "and while finding things to question in each one of them, I nevertheless learned a great deal from their study."[7] Reading Walter Rauschenbusch's *Christianity and the Social Crisis* he said, "left an indelible imprint on my thinking by giving me a theological basis for the social concerns which had already grown up in me as a result of my early experiences."[8]

King continued further, "One Sunday afternoon I traveled to Philadelphia to hear a sermon by Dr. Mordecai Johnson… he spoke of the life and teachings of Mahatma Gandhi. His message was so profound and electrifying that I left that meeting and bought a half-dozen books on Gandhi's life and works."[8] "It was in this Gandhian emphasis on love and nonviolence that I discovered the method for social reform that I had been seeking."[9]

Have you ever had to deal with an instance of complete injustice in the fire service or in your life? What can help you get through such an instance without being derailed in your progress? Do you know that when Jackie Robinson signed his first contract, it stipulated that he was not allowed to complain? So, if people called him racial slurs, which they did—a lot— or even when they spat on him, he not only could not react, he could not complain. Why did he allow this in his contract and put up with such indignities? He was looking at something much bigger than himself, as he talks about in his book, *I Never Had it Made*.[10] Read it and learn from him. Apply what he learned in your own life.

Unlike most jobs, in the fire service we have a good chance of facing a personal tragedy. Many firefighters lose their lives or are significantly injured every year. What if we lose a friend on the job? How do we deal with the tragedies we face, caring for our patients every day? What if we are injured to the point that we can no longer function as a firefighter? How do *you* deal with tragedies? We will talk about the importance of dealing with PTSD later in this book, but is there anything you can learn from how others have dealt with tragedy?

After the murder of his brother President John F. Kennedy, Robert Kennedy turned to the ancient Greek writers for comfort.[11] It helped him cope with an incredibly painful tragedy and helped him come to grips with the hate in humanity. It helped steer him away from the bad path of that hate. The writings not only comforted him but inspired him.

My own readings inspired me not only to continue to reach out to help others but also to get over painful events in my life. Can doing so help you? I believe it can.

Do you face issues that have surfaced from childhood difficulties or traumas? How do you prevent them from hindering your progress? Do you know about Oprah Winfrey's uphill battles since her childhood? How she overcame them and what led her to do amazing things in her life? Read about her, use her life as inspiration to endure the trials, and to not be derailed by the haters or the jealous people or the pitfalls and problems in your life.

Work hard to become the best firefighter or officer or chief—whatever position you are in. Take advantage of the incredible help available. Let others mentor you, personally or even by their writings or life experience.

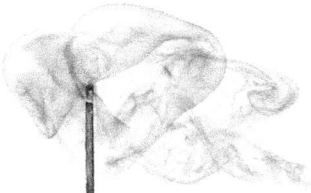

Chapter 13
BIRDS OF A FEATHER

Perhaps you have heard the saying "Birds of a feather flock together." In essence it means that people tend to associate with people like them or with the same tastes and interests. It's been used to suggest that you can tell a lot about a person and what they are like by his or her associates. That may be true, but we're going to look at this saying through a different lens, the lens of influence. The people you associate with and surround yourself with will influence your behavior and how you think.

So, you want the feathers of the birds around you to be those of positive, hardworking, movers and shakers who are progressing to better themselves. In fact, you want them to be better than you. Why? Because their actions and behaviors will rub off on you, *they will influence you,* they will make you better. And the opposite is true as well: associating with negative people will also rub off on you, influence you and make you worse.

Let me give you a personal example. I remember growing up on the streets of Brooklyn, New York. We played sports every day, in every season, and not just baseball, football, hockey, and basketball, but every variety of each sport, like stick ball, whiffle ball, frisbee football, roller-skating hockey, and more. It was awesome. I loved it and I loved sports. There were always kids older than me and there were always kids better than me. At first, when I was really young, I would get stuck playing against the best kids because no one wanted to play against them.

No one wanted to play against them because they were better. But what I found was that playing against the better kids started to make *me* better. I had to try harder. I learned tricks and moves from them. From then on, I always paired myself against whoever was the best on the other team. It really helped me improve. To this day I believe that is one of the things that helped me excel in sports.

The same thing holds true in every endeavor you work at in life. If you surround yourself with people who want to excel, doing things the right way, it will help you to excel. You will learn from them: their talent, attitude, way of seeing and handling things, their positives. These attributes and others will teach you, better you, rub off on you, and influence you.

It is essential to be aware of the influence on you from your subordinates, your peers, your superiors, and your own family. Even if you are in charge, as a supervisor, the words and attitudes of those you lead can have a major effect on you and your career. Author and motivational speaker Jim Rohn states, "You are the average of the five people you spend the most time with."[1] How do you add up? Let me tell you about a time when I was influenced in a negative way.

As I mentioned earlier, it took me close to three years to get hired by a fire department. Taking test after test for dozens of departments, with thousands of other applicants at times was morally defeating. The pressures of working full time and going to school full time were exhausting. Adding to the pressure, I wanted to marry my girlfriend, but I couldn't until I got this job. Sometimes I felt like I was never going to get hired.

When I finally did it, as you might imagine, it was a dream come true—the best job in the world. I loved it. I absolutely loved it! I would sit up at night daydreaming about it. I drove my crew crazy as I would constantly repeat, over and over and over again, "I can't believe I am a firefighter." I felt like a superhero! (That feeling never went away. Every time I put my bunker gear on, I felt like superman.)

When my chief gave me my badge my first week, he said to me, "Remember this feeling you have today, you will need it to carry you through in the future." I didn't understand what he meant then, but I would recall those words several times in my career.

The dining table in the fire department can be a dichotomy. On the one end, it's where we nourish ourselves for the next big event, where we share food together as a family. In almost a spiritual way it draws us closer to each other, creates a comradery, a brother and sisterhood with a bond that will never be broken, for many, until death. But there is a downside to the table. It's a breeding ground for gossip, rumors, negativity, and complaining.

The time during shift changes, over breakfast, lunch, dinner—you name it—can be used to pass rumors and complaints on, and they spread and grow like gangrene. And, like gangrene, they can and will kill. What will they kill? Your attitude, your motivation, your career. I have seen it. It is a fact! I have lived it. Within three years after the euphoria of my hiring, I was complaining about my job, the chiefs, everything.

One day, after complaining about things to my wife, she said, "John, I thought you loved your job." She reminded me, "You know you do have a really good job." I dismissed her comments, thinking she doesn't know what I go through. But after a little contemplation, I realized she was right. (This would not be the only time she would adjust my viewpoint and help me out.) I began to avoid talking to others when they were negative. But that wasn't enough, so I stopped coming to the table at shift change. I would stay in the dorm until close to shift change and would then meet the firefighter relieving me at the trucks. When coming on duty, I would relieve the firefighter immediately and then start my work. This proved to be monumental. I felt my love for the job coming back. I began to see past the negativities.

Avoiding the negative people can help you maintain a positive attitude, but there is something you can do that is even better: Associate with positive people. Surround yourself with them. To illustrate this, if you want to get in great shape, avoiding bad foods will help you maintain or even lose weight, but actively seeking out good meals and exercising will get you into great shape.

Positive people will help you grow. Negative people will draw you down, ruin you.

Dr. David McClelland, from the Department of Psychology at Harvard University, found after 25 years of research that the choice of a negative "reference group" was in itself enough to condemn a person to failure and underachievement in life. Your reference groups are the people you identify with—the ones you work with, socialize with, live with, and get involved with in community or non-work activities. Like a chameleon, you unconsciously adopt the attitudes, behavior, and opinions of the people with whom you most closely associate.[2]

Whether you like it not, those around you will affect you. Misery does indeed love company. Negative people will latch onto you to bring you with them. If you maintain your positive viewpoint and attitude, they will leave you, and positive people will be drawn to you. Birds of a feather do flock together. Ask yourself, what do my associates say about me? Are my associates known as positive or negative people in the department? Are they known as lazy or "ate up" with the job?

If you associate with the negative, lazy complainers, they will ruin your potential. Their effect will stymie your career. On the other hand, *positive people will nourish your motivation, they will feed your drive to succeed, they will lift you when you are down and push you forward.* Mark Twain wrote, "Keep away from people who try to belittle your ambitions. Small people always do that, but the really great ones make you feel that you, too, can become great."[3] Associate with the right people and you will not only feel great you will be motivated to be great.

How do you get started?

1. **Set the example**. Do as I do, not just as I say. Your example sets the path for others—friends, family, subordinates, bosses—to follow. You must live the right example.
2. **Always be positive**. The only way for other positive people to see that you are a bird of the same feather is for you to always be positive, especially in situations where no one expects you be so.
3. **Make it your core belief**. This must be your core belief, the core YOU because you can't fake it. If it's not your core, it will become visible. It will not have a positive effect on those around you, but instead, people will lose trust in you. Further, what positivity you displayed will be viewed as a cheap way to advance.
4. **Project this positive attitude in all you do.**
5. **Do not allow poison to be spewed at the table or anywhere**. Hold your crew responsible when they go negative, give them reason to be positive, help them to see the greatness in what they do and the good side—or even the tolerable side—of the things that are flushed down from up top.

Try to help your crew to always remember the love for the job they initially had. After all, we do have the greatest job on Earth. We are heroes. We save human lives! We are the ones people run to when they are afraid, when they are in need, when their lives are in trouble and we rescue them. What is better than that? Love what you do, love it and help those around you to love it as well. Then you will excel, you will help them to excel, and they will help you.

Every year around the month of May, I have memories of when I first got hired. The longer days return, the sun shines and the days warm. The city I worked for begins to empty out and the snowbirds begin to go back to their northern homes. Each year, it brings back memories of that 21-year-old boy who just got the dream-of-a-lifetime job. What a blessing. What a blessed life.

Gratitude!!

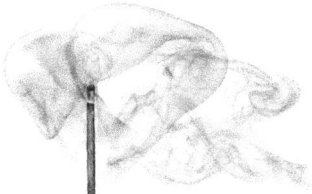

Chapter 14
KEEP CHECKING THE MIRROR

This book is about leadership, and the only way to truly achieving it is by starting with ourselves: doing a self-examination against principles that have always worked for leaders and addressing any deficiencies we find. It's a continuous process that should take us to the end of our career.

When I was on probation as a new hiree, I was put through an exercise where I was in bunker gear and a blacked-out mask so I couldn't see. We went to a large open field and I was supposed to crawl from traffic cone to traffic cone arranged in a plus formation. However, because I couldn't see anything, if I was even slightly off, I would miss the target and be headed to nowhere. This illustrates a point for us. If we don't keep checking where we are heading, we could end up off to nowhere.

As I was coming up through the ranks, I worked under two different officers who were constantly talking about leadership. They read books, went to seminars, and spoke to others in the department about these skills. The problem was, the two of them couldn't possibly have been more different. One was an extreme micromanager who oversaw every minute detail that went on with his subordinates. He was suffocating. In addition, he did not care for the people he was in charge of and he gladly told them at every opportunity, "I'm not here to make friends." He was a very difficult individual to work for and he was very disliked.

The other officer was the opposite. He despised any authority and worked to dismantle it throughout the department. He felt we all had to be just friends, except, of course, we had to follow his authority. If one of his friends worked with another officer, he did everything he could to undermine that officer. His friends didn't have to listen to other officers or do their station duties or even train.

I was perplexed as to how these two both thought they were the epitome of leadership. One drove people to do whatever they could to get away from him, the other drove people to think they no longer had to function like firefighters. Both would do much to hurt those who were assigned to them.

It was obvious that neither of them stopped to look in the mirror and examine what type of leader they were. They read and studied leadership, and both believed they were the best, and were right in the ways that they managed people. An Indian saying goes that books are as useful to a stupid person as a mirror is to a blind person. They had the information at their hands, *but they never honestly examined themselves against it.*

When we examine ourselves in the mirror, it's not a quick glance. We stare at our reflection to see the flaws we want to fix. None of us, looking in a mirror before going out, would leave dirt on our face. We would fix the flaw. The same should be true with our leadership.

How can we examine our leadership when we look in the mirror? Ask yourself reflective questions. Am I doing everything I can to be a great leader? Do I follow the department and lead others in that direction? Are my subordinates hurt by my actions? Do they love the fire service and working with me? Do they do everything they can to excel and become better firefighters? Do they take pride in the department? Are they good followers, not just of me, but of the other officers in the department? If I asked them what deficiencies they see in my leadership, what would they say?

Essentially, it's a process of examining ourselves against the result we put out, *continually.* Anne Frank put it this way in her diary: "How noble and good everyone could be if, at the end of each day, they were to review their own behavior and weigh up the rights and wrongs. They would automatically try to be better at the start of each new day and, after a while, would certainly accomplish a great deal. Everyone is welcome to this prescription; it costs nothing and is definitely useful. Those who don't know will have to find out by experience that 'a quiet conscience gives you strength!'"[1]

No matter how long or how short a time you have been an officer, keep checking the mirror to make sure you are not off on the road to nowhere. Make the needed corrections.

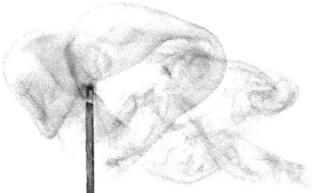

Chapter 15
NEVER STOP IMPROVING

Every person should live by this mantra, and certainly every firefighter should, because of the nature of our job. If you are a leader, it should be tattooed on your soul. We can never stop learning, nor should we. The moment we do, we begin to go backwards, we become a detriment to the department rather than an asset.

If you are the leader, more is expected from you and you should expect more from yourself. Henry Ford wrote in his autobiography, "I do not believe a man can ever leave his business. He ought to think of it by day and dream of it by night. It is nice to plan to do one's work in office hours, to take up the work in the next morning, to drop it in the evening and not have a care until the next morning. It is perfectly possible to do that if one is so constituted as to be willing through all of his life to accept direction, to be an employee, possibly a responsible employee, but not a director or manager of anything…but if he intends to go forward and do anything, the whistle is only a signal to start thinking over the day's work in order to discover how it might be done better."[1] Obviously, this thought applies to all—men and women. It's something that should constantly be with us.

In firefighting, we have a different type of job than most. When we don't perform as we should, people can get hurt or die. And when we lead, the same holds true for our subordinates. Shake off any laziness. The late, great Kobe Bryant once said, "I can't relate to lazy people. We don't speak the same language. I don't understand you. I don't want to understand you."[2] If you want to be a good leader you cannot be lazy. Keep working hard on the things that make you better. Confucius said, "By nature, men are nearly alike; by practice, they get to be wide apart."[3] Working hard will separate you from the others.

While each chapter in this book is intended as a guide to help you reach toward great leadership, encourage you to work hard, and get you started examining your performance in different areas so that you can improve—*don't stop here!* Continue to read books, go to classes and seminars, watch videos, and work with mentors. Never stop improving and you will be successful.

PART 3

WORK
HARD

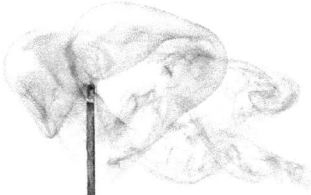

Chapter 16
ALWAYS SET THE EXAMPLE, ALWAYS DO WHAT'S RIGHT

As Will Durant stated, summarizing Aristotle, "Excellence is an art won by training and habitation. We do not act rightly because we have virtue or excellence, but we rather have these because we have acted rightly (…) We are what we repeatedly do. Excellence, then, is not an act but a habit."[1]

In the fire service, our ultimate goal is excellence and it isn't difficult to see the importance of doing what is right in order to achieve such. However, the importance here is on "always" doing what's right. Of course, none of us can "always" do anything, let alone the right thing. However, the idea is to *always* try. Keep trying. Never give up, never stop.

As we discussed earlier, habits, good or bad, right or wrong, are formed by doing the same thing consistently. Aristotle's insight into human beings is that when we make doing what is right a habit, we gain excellence and virtue in return. The depth of this concept cannot be underestimated. It affects everything around us and everyone around us: what they think of us, how they follow us, and what they will say about us. So, let's break it down into four points.

1 - Always set the example and do the right thing. How?

Doing what is right must be a habit. If you do something habitually it becomes second nature, the core of who you are, and it makes it easier to continue to do so. This starts with self-examination. You will not always do the right thing if you have character flaws, so you must correct them where they exist.

Do a self-examination. Examine who you are at the core level. If you find major flaws you must root them out and become a better person. If you live each day by the golden rule and do unto others as you would want done to you, then you will succeed. Examine your attitudes, your decisions, your thoughts, and your actions.

Never judge others. If you judge people based on their race, color, nationality, gender, religion, or sexual preference, you will not succeed. Human nature makes it easy for us to see the flaws in others, and make them bigger than they are, and not see the flaws in ourselves or to minimize them.

Care about those you lead. If you really care about those you lead, you will always try to do right by them. If you have a partner or children, you know that your care for them leads you to always try to help them, protect them, and guide them in the right direction. You care for them more than yourself. If you work to treat subordinates this way, you will always try your hardest to do the right thing.

Apologize when you make a mistake. Part of always doing the right thing is recognizing when you make a mistake and saying so. I have made many mistakes and sometimes big ones. Working in the fire service makes you part of a family. You work together for 24 hours at a time. You eat together, sleep in a dorm together, work out together, and at times just hang out together waiting on calls. You get to know people intimately. If you were raised in a big family you know that with all that close interaction, sometimes you say and do things you are not proud of. When you do make those mistakes, take action.

1. Admit it to yourself and whoever you hurt.
2. Apologize sincerely and never make an excuse for your failure.
3. Try to never make that mistake again.

When you apologize for your mistakes and work to fix them, it's like putting water on a fire. It completely stops further damage and gives you an opportunity to begin repairing the damage that was done. In addition, it shows humility, which is a necessary character for leadership, and you will also earn a reputation for doing the right thing. People will respect you.

Examine your reputation. What is your reputation? Ask others. Your reputation can draw people to you or make them not want to be around you. It can make subordinates want to listen to you or fight you on orders. It can aid you in promotional opportunities or hurt you. When you always do what is right, you are defined by your actions, they speak for themselves. Here is an example of a time when I saw the power of doing the right thing and building a reputation for such firsthand:

During one weekend shift, I had not yet been promoted but I was stepping up to engine lieutenant. We had a step-up lieutenant on the rescue unit that day as well. The station consisted of two units, with three personnel and one officer per unit. We had a new firefighter working with us that day, her name was Jessica[1]. Jessica was not assigned to our shift but that day she was working with us. She had been in the department for only a few months, but she was quite outspoken, even on probation. She was never one to muzzle her thoughts but instead would tell you exactly what she felt, whether you liked it or not.

The day started like any other, with us putting our gear on the engine, checking the unit, then going upstairs, doing station duties, and handling

---

1   Name has been changed.

the other prescribed items for the day. During the station duties, Jessica came up to me in front of a few others and said, "I heard that I need to watch out for you, that the first minute I'm not looking, you will stab me in the back."

Of course, I was stunned. I asked her, "Who said that to you?"

She replied, "I can't tell you, but is it true?"

I thought for a moment, immediately sizing up the crew and said the following: "I tell you what, Jessica, if I tell you that it is not true, then it's my word versus the word of whoever told you that, which will mean nothing to you, you won't know for sure. Of course, it isn't true, but I am going to let you figure that out for yourself. You take the next few years and then let me know if I am that person."

She was stunned by my answer and I went about my day as usual. I was highly disappointed. This type of childish nonsense is so destructive in the fire service. I felt confident it was the other acting officer who said it. What a terrible example. I had seen this behavior before. Years later, Jessica and I happen to be working together one shift. She came up to me and asked, "Hey do you remember what I said to you the first time I met you?"

I burst out laughing and replied, "Yes. Man, you are one outspoken individual." I then asked, "Did you figure it out?"

She said, "Yes and it didn't take me long, that piece of @#$* stabbed me right in the back. Do you want to know who it was?"

I laughed again and said, "I am sorry he did that; I am pretty sure I know who it was, it was Fred[2]."

She asked, "How did you know?"

I looked at her and said, "Jessica, everyone's true colors shine brightly. Do you think you and I were the only ones he's stabbed?"

Jessica and I would go on to have an excellent working relationship and she came to me for advice many times over the course of her career.

Work on yourself and your actions will speak for you. You do not need to say a word, just try to do the right thing and your example will shine.

---

2 Name has been changed.

2 - The effect that doing the right thing has on you

As noted throughout this book, any quality that you work on to make yourself a better leader will make you a better person in life as well—you will benefit from it. When the acting lieutenant, from the previously mentioned experience, said that about me behind my back, it really hurt my feelings. But unfortunately, it's the nature of the world we live in right now. However, Jessica came up to me years later and expressed that she saw and admired who I really was, that was a huge benefit, it had a tremendous effect.

It made me feel great about how I was living my life and the example I was setting. In addition, it felt great that people saw who I was, how I treated others, appreciated the way I conducted myself, and were drawn to me. Even more important, I knew I'd had a positive effect on them. It still makes me feel great when I think about that experience and it encourages me to always stay the course. Like a gardener who works hard in order to achieve a beautiful yield, you will experience the same. The gardener tills the land, plants the seeds, waters the plants, and weeds the fields. This hard work in due time creates a wonderful output and they enjoy the fruits of their labor.

The "fruits of your labor" that you receive *will reinforce your determination* to continue to work hard to always do the right thing. It becomes your identity. It becomes who you are, and that reinforces your behavior. To illustrate this point, if you know even a little about football, you know that Tom Brady is one of the greatest quarterbacks of all time and he continually shows us why he's great. Interestingly, as I am editing this chapter, he just won his seventh Super Bowl.[2]

However, even Brady has bad games from time to time. But when he has a bad game, he works harder the next week and comes out stronger the next time. It's not because he is trying to impress you, it's because Tom Brady knows who he is on the inside, he knows his identity. He knows that he is one of the best, that he is capable of greatness on the field, and he doesn't want to let *himself* down. It has nothing to do with others.

Likewise, when you get returns on your effort, you will begin to build who you are inside, and you will expect greatness from yourself. You will expect yourself to do the right thing always. When you fail, you will work hard to make sure that does not happen again because that is not who you are at the core. *You will act in accordance with who you believe you are* and what you think your identity is. Now imagine these fruits of labor not only in your leadership at the fire department, but also with all your relationships, with your partner, your children, your friends, your siblings, your parents. If you are always trying to do the right thing, it cannot help but carry over because that is who you are on the inside now. How do you think all those people just mentioned will feel about you?

3 - If you want to be a good leader, be a good follower

This statement seems oxymoronic. The theme of this chapter is "always set the example, always do the right thing." Obviously, a leader wants the people that he or she leads to follow. Thus, when you set the example, it's a pattern to be imitated. So, being a good follower to your superiors sets an example, a model to follow for those you lead. If you are not good at following your superior, you should not expect those you lead to be good at or even want to follow you.

But what if you feel your boss is not a good leader? This gives you a fantastic opportunity to show that you are a great leader. Over the course of your career in the fire service, you will encounter directives, orders, and rules you don't agree with. Sometimes they will be unpopular with the whole crew. You might have to enforce them anyway, even if you disagree with them. This is a challenging part of the job, but even if you disagree, the crew should never know.

We have all heard managers say things like, "I don't like it any more than you do," "they don't know what they are doing up top," "do what you have to do to get by," "play the game," "just follow it when they are around." It might feel nice, in the moment, to side with your crew and cast a negative light on the administration or the directive itself. But in the long-term, it sets a bad example. How do you think they'll respond when they don't like *your* directives? Will they just "play the game," or "do what they have to do to get by," or say that you don't know what you're doing behind your back? Shouldn't you expect that if that's what you taught them?

However, if they see that you accept unpopular directives, do not cast them in a negative light, and obey them whether your boss is around or not, they will respect you more as a leader. People will not think less of you for doing so, and they will learn from your example. People are not stupid, they recognize that you do not appreciate the directive either, without your even saying it. Your humility in accepting a difficult directive, not pointing the finger but, instead, following orders, being a good follower, will have a great effect on those you lead. These actions will build their confidence in you.

They will be more willing to follow your directives or orders in the future. In addition, understand that when you deviate from the administration that makes the department run, when you cast their directives in a bad light, you are encouraging dissention. When you do this, you hurt the whole organization and you hurt the subordinate as well. How? Well, no one wants to feel like they are being led by bozos or people who don't care for them.

This is your department too. Do the things that make the organization great, that build it up in the eyes of all the firefighters. We all have to do things we don't like sometimes in this life of ours. Sometimes we don't understand the wisdom in it until later on.

Everyone in the fire service has a boss. Everyone. So, everyone is going to have an opportunity to show that they can be a good follower.

Are there ever times to not be a good follower? What if you feel a directive is dangerous or you just don't understand it? As a good subordinate, you should discuss this with your boss, but do it privately, not in front of the crew. In addition, there may be times when you have to flatly disobey orders and take the punishment. If your boss tries to get you to discriminate against someone, break department polices and rules, put someone in harm's way, hurt someone physically or mentally, or hurt their career, you should not participate in it.

4 - Your effect on others

We spoke earlier about how you are the average of the five people you surround yourself with, how those around you have an influence on you, your behavior, your thoughts and actions. Well, the same holds true in reverse. You have an influence upon the people you are around as well. Your actions, thoughts, and behaviors influence them. Your example of trying to always do the right thing will impact them to do the same. Over time, they will see the benefits of your example. Once they move in that direction and taste how good it makes them feel, they will be moved to continue to progress.

As they begin to refine themselves, your impact will grow as others are influenced by *them*. I remember a shampoo commercial from my youth where the girl says: "And I told two friends and they told two friends and so on and so on." As she was saying those words, the screen was splitting with pictures of more and more people. The exponential growth was impressive. While you may not see the exponential growth of your example, you will certainly have a significant impact, by virtue of your example, even upon people you had no interactions with.

Know, too, that one of your top priorities in leading others should be not only to get the job done right but to help create the next generation of leaders. You have a duty and an obligation to your department and those you serve to help develop the next generation of leaders. Why wouldn't you want the best leaders possible? At all times, you should be trying to make the next person coming up a better leader than you are, to learn from your mistakes, and repeat your successes. In doing so, your example in the department will be felt for generations.

We also have a duty and an obligation to our country and as a citizen of the Earth to help teach others to always do the right thing. This is a quality we want our police officers to have, our teachers, our politicians, our bankers, our doctors, everyone. So, if you want this from others, if their being this way will make the Earth a better place, then always set the example, always do the right thing.

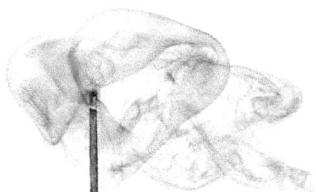

Chapter 17
EMBODY WHO YOU WANT TO BE

We spoke about the journey we're on and it being a road to becoming a great leader. This book is full of guideposts and ideas on becoming that leader, but just as in a vehicle on a road, unless you hit the gas pedal, you will not go anywhere even with the best directions. You can read every leadership book written, you can work with one of the great leaders in history or have them as a mentor, you can have all the desire in the world to be great, but unless you move forward and apply these lessons, you will never become that leader yourself.

How do you hit the gas pedal? You take *full personal responsibility for yourself and your actions* toward embodying what you want to become. It takes more than acquired knowledge. You must eat, sleep, and live it. It must change you as a person so that everything that you do is affected by the changed you.

The American novelist Nathaniel Hawthorne wrote in *The Scarlet Letter*, "No man, for any considerable period, can wear one face to himself and another to the multitude, without finally getting bewildered as to which may be true."[1] It cannot be a face you put on, a mask, an act, a trial, or a lack luster effort. If you approach it that way, you will fail, and you will actually be in a spot worse than before. Not because you failed—failure is both inevitable and necessary—but because those whom you wish to lead will see right through your lack of conviction and will lose respect for you, knowing you were putting on a show. In addition, they will lose trust in you, second guessing anything you do or say.

Early on in my career, I worked with an officer I became quite fond of. Unfortunately, as timed passed, he put less and less effort into being a leader. The last few years of his career, he would not even dress out on fire calls, many times leaving the firefighter to join with another crew on an emergency scene. His lackluster effort was so well known that the department firefighters began to refer to him by a nickname that highlighted how little he contributed. The name was quite funny but very demeaning. The FF's no longer had any respect for him. It was so sad to see. I was sad for him, a man I liked, and I was sad for my department.

When you give something of extreme importance only a halfhearted attempt, it shows a lack of respect for yourself and a lack of discipline. If you don't respect yourself, no one else will respect you. When you are in a position to lead others, and you have the lives of patients in your hands, you cannot give a half-hearted attempt; it shows a lack of respect for the person you are caring for as well.

"All change is from the inner to the outer. All change begins in the self-concept. You must become the person you want to be on the inside before you see the appearance of this person on the outside." Brian Tracy made that statement in his book, *Maximum Achievement*.[2] The thought of changing ourselves is very difficult. In fact, it could be one of the most difficult jobs that we can do. That is why most people don't do it or they try it and eventually quit. But if you have the courage, if you have the strength, if you have the fortitude and endurance to do so, you will achieve whatever it is you want. In fact, you will be able to achieve goals in any area. A person's worst enemy is himself. If you can conquer yourself, you can conquer anything, you will be unstoppable. And when you accomplish that, others will be drawn to you. They will want to be in your circle, to learn from you, and be led by you.

When I read Chris Kyle's book, *American Sniper*, I understood something about his life that, in my opinion, is what made him so special at his job as a Navy SEAL. He explained that ever since he was a young boy, he'd always wanted to be a cowboy, and how he had been content living the life of a cowboy. But I noticed something, when circumstances changed in his life and he found himself enrolled in training to become a Navy SEAL—one of the most difficult training camps there is and a profession he'd barely heard anything about—it began to embody him.

As I read his story, I realized his new career became his blood, who he is, his identity. He continually stated, "I am a SEAL down to my soul," and "once a SEAL always a SEAL."[3] Over and over again he referred to himself that way—not as a man, not as a cowboy, but a SEAL, even after he retired. I understood this because that is how I feel. I was an electrician's helper in high school and continued the training and work when I got out of high school. But then a recession hit and changed circumstances which forced me to look for different employment.

I searched out and placed my application all over for a more stable government-type job, and the fire department was one of those applications. I obviously knew about firefighters, but I didn't know much. I had no idea what the work was like other than putting out fires, nor had I ever had the desire to be a firefighter. But when I went to the Fire Academy, I was hooked. It began to infiltrate my blood. It became who I am. It became my identity and I stood proud throughout my career. No matter what professionals were in the room, when asked what I did, I was proud to say, "I am a firefighter." And I will be until I die. I love the fire service.

There are many roadblocks when you hit the gas pedal and start moving forward. These can surface and even resurface somewhere down the line once you've already begun to pick up speed. You can categorize them by three major categories:

1 - Ignorance

If you do not know how to lead others, have some understanding of how people work, or how to do your job, you can't possibly lead. The fire service is one of a few jobs where ignorance costs lives. We can't afford not to be knowledgeable. Fortunately, this is the easiest area to work on, but easy doesn't mean quick. How do we attain that knowledge? There are college classes that help, books, seminars, videos, and a host of other tools that you can turn to.

What books are you reading right now, besides this one? How many have you read this year? You should never stop reading and learning.

If you don't like reading, try starting with a book you *think* you might really like, perhaps a work of fiction. Then move on to something to help you with your goals—or listen to audiobooks. Never stop. Learn from a mentor as well. Ask specific questions about how they do things and watch how they handle situations.

When I responded to emergency calls as a firefighter, I would determine in my mind how I would handle the situation if I was the officer. Then I would watch how the officer actually did it and I'd ask questions afterward, especially if what they did was very different from what I had thought. No matter how much you learn, you will still need to continue learning once you get into next position.

2 - Fear

It takes tremendous courage to be a firefighter, choosing to run into chaos and destruction, risking our life for others. We cannot show fear in the face of emergencies. And we don't!

There is another kind of fear, however, that can prevent us from fulfilling our potential. Many are familiar with President Franklin Roosevelt's words, "The only thing we have to fear is fear itself."[4] Do you know what he was talking about? His next line would explain the detriment of fear. He continued: "Nameless, unreasoning, unjustified terror, which paralyzes needed efforts to convert retreat into advance." Whenever people are asked to do something new or big or different or to put themselves out there, fear develops. That fear can paralyze them.

They are afraid they will fail. They are afraid they won't look good in front of others, they will be made fun of, criticized, laughed at. First, let me put you at ease. You will fail. And you will fail a lot. Failure is good, as we've discussed, it's necessary.

Second, when you fail, some people may think badly of you. Some people might mock you for trying to better yourself. They might criticize you and laugh at you. This is generally because you are reaching out, despite your fear and that makes them feel badly about themselves, perhaps because they are afraid themselves. You never want to let your own fears hold you back from moving forward, but certainly, unequivocally, never *ever* let someone else's fear hold you back. Conquer fear and nothing is unattainable for you.

3 - Hard work

Hard work is a necessity. I think that point is obvious. Let's talk about two specific things that cause us not to work as hard as we could and should.

First, we tend to work harder at things we are familiar with or have done before. For example, hose drills: Pull the hose off the engine, flake it out completely, charge it with water, advance and flow it, shut it down, break each individual 50-foot length of hose apart, drain them, reconnect them, stack the hose back in the engine, all while fully bunkered out in 95 degree temperatures only to do it all over again. It's extremely hard work. It's brutal. But you do it all the time. You have done it since the academy, and you can do it over and over again.

However, when we try something new, something that is very difficult, it's hard to put forth the continued effort and very easy to quit and go do something else. Learning something different makes you become something different than you are right now, and that is one of the most difficult challenges we can face. Though we work hard every day, we must be careful not to quit when we run into difficulties in tackling change and learning something new. Eventually, with the effort, you will master that new challenge—and it will become as familiar, and repeatable, though perhaps no less difficult than the hose drill.

Second, we have to be careful we don't fool ourselves about how hard we are working. It is easy to do. People say things like you have to give 110% or 150%, or they quit because they gave the job 200% and it didn't work out. It's not possible to give 110%, 150%, 200% or even 101% of yourself. I think people just use those numbers to trick themselves or give themselves an out. When I read David Goggins's book, *Can't Hurt Me*, I found a kindred spirit in this regard. Ultramarathon runner and former SEAL, Goggins speaks about something he calls the 40% rule.

He says that the human body is like a car in that it has a governor in it that will stop us when we begin to push too hard. But, unlike a car, we can push past the governor. Goggins further states, "Most of us give up when we've only given around 40% of our maximum effort. Even when we feel like we've reached our absolute limit, we still have 60% more to give!"[5]

Now, is he right with that exact number? Well, of course no one can be certain. However, I tend to agree with him, and if you get a chance to read his book and see where he came from and what incredible feats he has accomplished, the things he has been able to push himself to do, perhaps you may feel he is expert enough to name the percentages. And perhaps his story will help push you past your own 40%.

Why is this important to understand? Because when we tell ourselves that we gave 150%, we tell ourselves we gave it all and since we failed or it didn't work out, we are then telling ourselves there is nothing more we can do. However, when we accurately know we probably only gave 40%, 50%, or maybe 60%, then, if we should fail, we know we can give more the next time. When we know that, we will not only continue to keep that foot on the pedal, but we will press harder after each stumble or failure. We will not stop.

We can't give more than 100%. Giving that much of yourself to something is near impossible in the world today. Both the body and the mind can be pushed to such incredible feats, far beyond what we understand. When Joe Rogan was interviewing David Goggins and they spoke about this 40% rule, Rogan stated "100% is at death's door."[6] Goggins agreed, as do I. We want to push hard toward whatever we are working at but know we can almost always give more. Keep working harder.

So, in conclusion, when we strive toward something we want to be, we are reaching out to become someone different, a new identity. This will take great effort. It will take us learning new things, breaking through any fear we have and working hard. However, when we get there, when we become who we want to become, it will infiltrate our blood and our soul, it will become our identity and this will automatically make us want to live up to who we know we are.

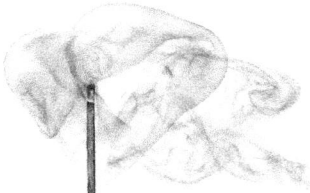

Chapter 18
USE YOUR TIME WISELY

This is an area where I felt like I was very successful, and yet my success presented some of my biggest challenges in the fire service. How could that be, you may ask? Let me explain a few of the challenges and tell you what I have learned from my experience and how you might perhaps face the challenges better or, at the very least, never let anyone stop your progress.

To start, you should know that I am an individual who has many friends. I enjoy the company of others and love to get to know new people. I am also a very driven individual, I'm focused, I love to work, and I both need and enjoy my alone time. Early on in my career as a firefighter, I was amazed at how much time we had during a shift (24 hours), in which we could do our own thing. Weekdays from 0730 to 1600 were busy with necessary chores in between calls, but after 1600 we could work out and after 1700 the day was ours, with the obvious exception of emergency calls and the things that came along with them. As time has passed in the fire service, I've watched as the schedule has changed drastically in South Florida and many other areas in the country.

We have taken on and absorbed the emergency medical field and significantly ratcheted up the training. Hazardous materials and terrorism have become much bigger focuses, school shootings and natural disasters as well, including things like pandemics. We are very busy while on duty. But by nature, when working 24 hours at a time, we have what the industry calls "down time," when we are at work but not working. Sometimes it might be only 10 minutes in a full day, but often it can be an hour or several hours. I've never been a fan of spending my time with TV or the internet. I haven't yet participated in social media either.

I occasionally watch football and hockey, but I can't even do that without working at the same time. Only *Fast Money* on CNBC gets my full attention, it's the best show on TV, in my humble opinion. Because of this, when the crews that I worked with watched TV, I would do my own thing. During the day, if there was down time, I would work on projects in the bays or find a quiet room for study or work. At night, I would go to the dorm or my room and use that time to read or study or do work on my computer. There were always new things to learn and get better at in the fire service.

In addition, I used my down time to study finance and financial markets. I was elected as the pension representative for the Police and Fire Departments in our city and served in that capacity for almost 10 years—which was a huge learning curve. I also organized a tremendous amount of fire and medical training for myself, my crews, and the whole department through the years, and built a number of training props. This included building out an old furniture warehouse with training scenarios and taking the whole department through them over the course of an entire summer. I did the same thing another year with an empty building that a neighboring city let me use.

I tell you these things only, which is the abridged version, so you can see that I had an enormous amount of interaction with all the firefighters and in many different ways. I was anything but a hermit. And yet, "my not watching TV with the crews" became an issue for a few of the officers I worked with.

It was used against me on a few of my evaluations. This was extremely puzzling to me and difficult for me to accept. There simply was no way to rectify the action of these officers and the impact it had upon my evaluations and subsequently on me and my family. It left me frustrated for so many years.

Interestingly, in reading David Goggins's *Can't Hurt Me*, I discovered that the same issue affected him. He tells how, while many others in his team liked to party at night, he didn't. He used the nights "to rest, recharge, and get my body and mind right for battle again the next day." He wrote that his OIC (Officer In Charge) tried to influence him to let go and become "one of the boys," telling him, "you would understand the job a little better if you hung out with the guys more."[1]

"His words were a reality check that hurt," Goggins wrote, but he recognized that the OIC was responsible for his evaluation and that "evaluations have a way of following you around and affecting your military career." Goggins had thought that his differences, his intensity, were the things that got him into the SEALs. "I thought that made me part of the group, but now I realized I was part of the Teams—not the brotherhood."[2]

WOW! I felt like I was reading my life story. I have lived this very scenario several times in my career. All the confusion on the topic, and the many strange evaluation conversations flooded my brain. From the first time my lieutenant told me I should be watching TV with the boys instead of reading and studying, to subsequent evaluations where it was pointed out to me that I don't eat breakfast with the guys, I don't associate enough outside with the guys, I shouldn't be reading the *Wall Street Journal* at work, I should chill out and not train so much, I shouldn't read in my room, I shouldn't study in my room, on and on and on.

> *The brotherhood that I so longed to be a part of became just a word, and then a weapon used against me.*

How could an officer want me to watch TV instead of train or study, and then use it against me? I was bewildered. When you enter the fire service, you walk into a job, new, green, looking up to those already making up this family, surrounded by what you believe are the bravest, most courageous men and women in our country. And the officers *are* leaders, full of knowledge and insight, ready to teach you, train you, and mold you to be a great firefighter.

But while this may be true in many instances, it isn't always the case. You will encounter FFs and officers who are insecure about themselves and ones who are incompetent. Some will not want you to become better because their own incompetence shines in their face. My initial interaction began to make me doubt that this family bond I'd heard about really existed.

In time, I came to realize that it most definitely does exist. Unfortunately, some will try to destroy it or manipulate you for their own means, but they don't own the fire service. You are a part of that family and you help create it every day. When you use your time wisely, you make yourself a better firefighter, a better officer, and a better member of the fire service. You begin to mold the fire service in a positive way, you set an excellent example, and your birds of a feather will start to come and flock together with you. Do not let anyone use the family bond to get you to slow down in your training and improvement of yourself.

Henry David Thoreau once wrote, "The mass of men live lives of quiet desperation."[3] For whatever reason, people feel a void in their lives. You have been given a great gift. You work in the fire service; your career is in the service of helping people. Follow the passion that brought you into the fire service, do not live a life of mediocrity, do not live small, do not let anyone slow you down.

The key to friendship in the fire service is this, it's important to try to develop those friendships, but it's more important to develop yourself. When you're working to be the best you can be, you will attract the right friends, the ones that will build you up, make you better, support you, be there for you when you need them, have your back in difficult times and so much more. Anyone who tells you that you should be watching TV with the group instead of bettering yourself, supervisor or not, does not have your best interest at heart, nor the fire service.

Here is where having goals, finding a good mentor, and having good associations will help you. Use your time wisely and you will achieve many great things, including your goals. And even if some individuals should be intimidated by your good use of your time, and in some way that affects you, in the long run, if you are persistent, they and their actions will be just a small blip on your wonderful journey in the fire service.

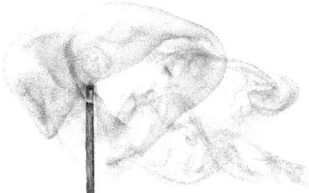

Chapter 19
LEARN TO LOVE DISCIPLINE

The word discipline may have a negative connotation in some people's minds. The level of self-discipline one has in his or her life, however, can be one of the greatest determiners of their success and happiness. Why should we learn to love discipline, especially in the fire service? Two reasons. First, loving what we do makes doing even truly difficult things easier, smoothing the edges. Second, discipline is perhaps the single greatest attribute to mold you, strengthen you, teach you, prepare you, and shape you. We'll discuss three facets of discipline in this chapter: self-discipline, receiving discipline from others, and finally, giving discipline.

Let's start with the most important: self-discipline. Dictionary.com defines it as, "training of oneself, usually for improvement."[1] That's a start, but it's a bit vague. Dr. Myles Munroe, a leadership consultant, puts it this way: "Discipline is self-imposed standards for the sake of a higher goal."[2] That helps us get a better grip on it, focusing on the end results. If we know what we are shooting for, it's much easier to work toward it, it's much easier to train ourselves or impose standards on ourselves that we try vigorously to live by. Let's look at an example.

If someone wanted to lose weight, they might say their end goal was to lose 25 pounds or to fit into a certain size, to get healthier for firefighter work, or just to feel better. That is their end goal, so they set up a self-discipline to achieve it. Perhaps they lower their caloric intake, exercise more, and eat more vegetables. Whatever it is, discipline comes in when they have to do that every single day.

Our end goal in reading this book is to become a great leader. So, we establish the things that help us get there and stick to that routine every day. Self-discipline, as the name suggests, starts with us individually. In fact, *discipline is self-leadership.* How can you lead others if you cannot lead yourself?

Part of self-discipline is finding out what areas we need to improve. This book and others outline many qualities that you can examine to see if and where you need improvement. Peers, supervisors, and trusted subordinates can also add to the list of qualities you may need to work on. Those who love you are an excellent source of inquiry about any deficiency they may see in you. Although that can sometimes be hard to hear from those you love, it can be the most accurate assessment since they know you the best.

All this might seem overwhelming, it can seem like too much work to constantly be disciplined, however for those who live disciplined lives, it becomes a part of them.

When someone starts working out for the first time, it's not easy. It's difficult to fight yourself to get to the gym or to go out and run every day or to limit your calories. Worse, you don't see results right away. So, you fight yourself to work out today, and then you do it again tomorrow, and then again the next day. A week passes and you do not see any results, perhaps you lose a pound or two, but that doesn't feel worth the amount of effort that's gone into it.

But the truth is, there are results. They may be small at first, but they build on each other. And something more important is happening: you are winning the battle with your mind over the effort to discipline yourself. It is becoming a little easier. At some point in the processes, a shift happens.

Eventually your mind gets into the routine and it isn't a fight to work out anymore.

In time, your mind will be so disciplined that it will fight you if you try to stop. This doesn't mean there will never be days when you are out of sync and don't feel like working out. There will be. But you will soon find that on those days you don't work out, you feel out of sync, like something is wrong. Then you'll feel excited to get back at it again.

That is the process with self-discipline. If you stick to it, you will win the battle. Exercise is just one example. The same holds true for any self-discipline tract you take. If your goal is to quit drinking soda, stop thinking negatively, or start reading more, you will have to work at it until it feels like second nature to you.

Do a self-examination against the qualities described in this book, the qualities needed for leadership. Ask your supervisors, your subordinates, and your loved ones how you measure up. Then make a list. Work on those qualities one at a time.

Some firefighters are hired at very young ages. They may finish high school and then go to the appropriate schooling for the fire service and get hired right away. It can sometimes present challenges for a young person, when their first real job is in a para-military organization, especially if they didn't expect the rigidity and discipline that goes with a fire-service career.

We hired such an individual. He was a fantastic young man, and since I too was young and still relatively new, we had a lot in common and became good friends. He had difficulties adjusting to the department. He was happy and jovial but had a less-than-serious view of his job. This attitude, along with a less-than-studious nature, caused him to not do so well. He failed a few probation tests and was on the brink of being fired. He approached his supervisors, almost in tears, asking what he needed to work on.

In a matter-of-fact manner, he was given a few things to work on and was told, essentially, this is it. Improve or you will be terminated. The self-discipline kicked in. He buckled down and worked really hard, focusing on one deficiency at a time. It was amazing to see him transform. It was like watching someone grow up overnight. For me and for everyone else who witnessed it, he set an incredible example of what self-discipline can achieve.

I heard actor Will Smith define self-discipline this way, "Self-discipline is self-love."[3] He is telling you that whatever difficulty you may encounter, being self-disciplined will ultimately benefit you. You will be happier for it. Discipline, in fact, may have a surprising outcome. Notice how it is emphasized by Jocko Willink and Leif Babin in their book, *Extreme Ownership, How U.S. Navy SEALs Lead and Win*. They stress that "Discipline equals freedom," and further explain that "Although discipline demands control and asceticism, it actually results in freedom."[4]

Willink adds, "I realized very quickly that discipline was not only the most important quality for an individual but also for a team." (He then described the team disciplines) "But there was, and is, a dichotomy in the strict discipline we follow. Instead of making us more rigid and unable to improvise, this discipline actually made us more flexible, more adaptable, and more efficient. It even allowed us to be creative." Seeing the benefits, he continued to apply self-discipline. He stated, "As I advanced into my leadership positions, I strived to constantly improve my personal discipline."[5]

How does discipline bring you freedom? In many ways. Specifically, in the fire service, when you are disciplined, you are good at what you do. You have a good knowledge and understanding of your job. You earned that through a disciplined approach to study and training. The young man mentioned earlier, who was doing badly on probation, became nervous every day about what he was doing. He was terrified he was going to make a mistake because he didn't know his job well. He was not confident. He said he felt like he was walking on eggshells, waiting to make a mistake and get fired. After working on his deficiencies in a disciplined way, he had a good understanding of what was expected of him. He became good at his job and he experienced the freedom of coming into work and not worrying, knowing that he knew what he was doing.

As a leader, being disciplined gives you freedoms on emergency scenes as well. No two calls are exactly the same, so it is hard to have an exact protocol on how to handle something. Knowing your job well allows you to improvise and try different tactics. It gives you the freedom to lead different personalities in different ways. Self-discipline can benefit you in all areas of your job and life.

Nelson Mandela recognized the power of self-discipline throughout his life, including while he was in college and competing in sports. He wrote in his book, *Long Walk to Freedom*, "In cross-country competition, training counted more than intrinsic ability, and I could compensate for a lack of natural aptitude with diligence and discipline. *I applied this in everything I did*. Even as a student, I saw many young men who had great natural ability, but who did not have the self-discipline and patience to build on it."[6] The self-discipline he learned and applied in cross-country running he used throughout his life for his personal success. The same will happen for you.

RECEIVING DISCIPLINE

This may be one of the most difficult experiences in the fire service, and it's not just difficult for the person on the receiving end. Receiving discipline can be a knock on our self-confidence and self-esteem, depending upon the severity. I have received discipline many times in my career. How should we receive it? Remember that it can be a significant asset for you in your career and life.

Discipline is a straightening out, a correcting of the course. Like a cast placed on a broken bone, discipline serves to protect and help set you on a straight course. If you receive it correctly, you'll never need it again in that area. Once, after a huge mistake I made in my career, I received strong discipline. I evaluated the situation. I saw how I went astray, and I fixed it, it changed my career, it changed me, it benefited me enormously.

But that does not mean I always took discipline well. Early in my career, I sometimes really chaffed at it. I hated it. But I learned how important it was, and in addition, I learned not to let it affect me negatively.

When you receive discipline, can you do an honest self-examination? Can you look at the discipline itself and not at who gave it to you, or how they delivered it? Did you do something wrong? If so, can you fix it? It's hard to filter out all the noise around a circumstance, but doing so will help you accept it and rectify it.

For example, imagine if you were swimming at the beach and a lifeguard started blowing his whistle and waving you in, right in front of everyone. Everyone was looking at you. Maybe you felt embarrassed. Then, as you got up to the shore and looked back at the water, you saw the massive fin of a shark right where you were. Would you care anymore that the lifeguard whistled and pointed at you in front of everyone and told you to get out of the water? He saved your life.

Can we accept discipline that way? Try to look back and see what it was that caused the need for discipline. If you can accept it and become better for it, you may be surprised by the reaction you get from those who gave it to you. Is it possible to look upon a discipline with gratitude? Someone cares enough to do the difficult thing, to help you. If you can accept it with grace, it will draw you closer to your officer.

When we get good at receiving discipline it also makes it much easier for us to give discipline. You understand how it feels, so you are less likely to make it worse for the person you have to correct. And when your subordinates see how well you have taken discipline, as an officer or a firefighter, it will make it easier for them to receive it from you as well.

GIVING DISCIPLINE

One of your duties as an officer is to give discipline when needed. As a leader you need to know when to step in and correct those you are leading. This activity may be one of the worst abused activities in the fire service, mostly out of lack of training and partly out of malevolence. I know that I personally failed many times when I was promoted. Initially, I wanted to be a well-liked officer, so I didn't discipline much. I was taken advantage of at times, but the people who lost most were those I didn't discipline when it was necessary. I failed to lead them and set them on a straighter course.

When I recognized my mistakes, I worked to change. Unfortunately, that led to times where I then failed by disciplining too often. I never received any training on how to correct subordinates. One day I was a driver/engineer, the very next day I was an officer. This isn't an excuse, it's just reality. After much study, thought, reflection, many mistakes, and discussions with other officers, I finally came to, what I believe, was a good balance.

I'll outline a few things to think about, but keep in mind that your department's protocols must always come first, so that is where you want to start. Additional support can be ascertained by asking a superior whom you have seen do an excellent job when it comes to correcting individuals in his or her care. Perhaps this superior can be a mentor. The individual should be well respected for being fair: not too soft and not too strict. Try to gain a good understanding of discipline's importance and why there is a need for it. Remember that it encompasses a full range of actions, from written reprimands to a simple conversation, which should be the discipline of choice in the far majority of cases.

REASONS FOR DISCIPLINE

The reason you give discipline is to correct someone's course of action. It is to help them and their career, not to hurt them. It is to help and protect the customers we serve, not to hurt them. It is to benefit those receiving the discipline and those served by them. Those things should be among your first thoughts when considering the need for corrective discipline.

WHEN TO DISCIPLINE

It should be done as soon as you see someone taking a step in the wrong direction—a simple lie, a simple rumor, not checking the equipment correctly, a wrong attitude. Heading off a problem with a simple corrective statement, question, or conversation can save that person much worse discipline in the future.

MORE THAN ONE HOW-TO

It can range from a simple verbal, "don't do that" to a one-on-one meeting for a deeper discussion, or even a discussion with a written reprimand. If the need arises for a written reprimand, someone failed along the way and more than likely it was you. When lines of communication are open, information is constantly flowing up and down the chain of command. This in itself, will help to eliminate the far majority of issues.

Always treat the person being disciplined with the utmost respect. Just because they did something wrong does not give you the right to be disrespectful. Protect their dignity as well and always do it in a way that is helpful, don't destroy their trust in you. If you genuinely care about the person and their career, you will always look for the best way to help.

Here's a situation that happened to me and what I learned from it in my third year at the department, as a 23-year-old kid. I was still so new and there was so much still to learn. My station duty entailed vacuuming the carpet in the dorms in the morning and sweeping and mopping the tile floor in the station in the evening. I hated that duty so much. While everyone else was done with their duties for the day, I still had work to do. Of course, I look at it now and think how pathetic it was that that little duty bothered me.

There was a section of the station, a hallway, approximately 3 feet by 7 feet that was almost never used. One night, I decided to skip mopping that small section. It looked clean enough, and no one ever used it. After I was done for the night, my lieutenant asked me if I had done my duties and I said "yes." I thought it was a strange question since he had never asked me that before.

During the next shift, he pulled me into the office and said that I'd lied to him. "When?" I asked. During the previous shift, I was told, I stated that I'd completed my duties when I really had not. I explained that I had not lied: he had asked me if I did my station duties and I replied "yes." I honestly felt like I *had* done my duty: the hallway was clean and never used so I didn't think I had to mop over it again and I did the rest of the entire floor. He repeated that it was a lie.

I disagreed, but I apologized: I told him I honestly didn't know that the hallway had to be mopped even if it was clean, and I promised I would do it from then on, no matter what. He did not write me up. That was the end of the conversation, or so I thought.

That afternoon the assistant chief came to the station like he did each afternoon. He spoke with the crew and then asked to see me in the office. I went in there with him alone. My officer didn't accompany me, which I thought was strange.

He immediately said, "I heard you lied to your officer."

"What?" I replied with utter confusion.

"I heard you lied to your officer," he repeated.

I was shocked. I asked, "Is that what Lieutenant Jones* told you?"

"Yes," he stated.

I explained my side. He told me he did not care; the point was that I lied. I disagreed with him but stated I wouldn't do my duties like that ever again.

I then asked him this question. "If Lieutenant Jones has to run to you with every little thing that happens with his subordinate here in the station, why do you need him as an officer?"

He seemed stunned. This was a man who was always angry and this meeting was no different. He reveled in disciplining others. He paused for what seemed like 10 minutes but was probably only about two. His angry countenance fell.

"I guess you're right, that's a good point," he said as he stood up. As he walked out, he said, "Don't ever lie again."

I mumbled, "I didn't lie."

Here is where failures happened. First and foremost, let me say that if I had done all the mopping I was supposed to do, we would never have had the problem. Second, if I had said in the beginning, "I did my duty but skipped the little hallway because it was clean already," I wouldn't have had that problem either. But that doesn't solve all the problems because people and situations always vary. We live together for 24 hours at a time at the station, eat, drink, sleep, workout, train, run emergencies, and do hundreds of things together, and no one is going to know exactly how everything is expected to be done by each and every officer. People are raised differently and see things differently. In addition, working in the fire service is a completely new experience for most of the people who start there. The point is, there will be some confusion for newer firefighters learning how the service works.

This officer asked me if I did my duties, and I said "yes." He obviously knew I hadn't done the small hallway, otherwise he wouldn't have asked me the question. He didn't like that I hadn't done the hallway and he felt I was dishonest with him.

1. The very next step should have been, "let's talk in the office about this." **He failed to do that and waited until the next shift. (When he finally approached me, this left me confused about his leadership skills and feeling that he was weak.)

2. He told me, the next shift, that he knew I did not do the hallway. **He placed me on the defensive and now it appeared to me as if he was trying to trap me. (At this point I felt like I didn't trust his words. He purposely tried to trap me in order to make me look bad. Would he do that again in the future?)

3. He immediately told me I lied to him, I was dishonest. **He accused me, tried me, and convicted me of something very bad, something I hate, without even asking me for any information. (I then felt I knew he couldn't be trusted and nothing I could say to him would help. He had already convicted me.)

4. Demanding an answer, I explained to him that because the hallway was already clean, I didn't think it needed to be cleaned again. That was not sufficient, he stated. **He again repeated his accusation. (Now I confirmed to myself that he just wanted to make me look bad, he did not really care.)

5. I apologized profusely because I genuinely didn't want to do anything wrong and I prided myself on being honest, which I explained to him. I stated that from then on, I would mop everything no matter what. He still insisted I wasn't being honest. (At that point I was boiling with anger inside, but I accepted it and left. I had lost respect for his leadership and I didn't trust him anymore.)

6. Then the assistant chief pulled me into the office. **The officer further escalated this issue. (I was completely stunned that my officer had told the assistant chief. I lost even more respect for my officer, feeling he was completely weak. I now knew that I would have to be careful about everything in front of him, I could never *ever* trust him.)

7. The assistant chief repeated the accusation. **Now he too had accused, tried, and convicted me of being dishonest without even a word from me. (I was exasperated, bewildered that this was really happening over my not mopping a section of flooring. I now placed the assistant chief in the same category as my officer. In addition, the assistant chief thought negatively of me because of something he should never have heard about. It should have been worked out with my immediate supervisor.)

In the end, what could have been an excellent teaching moment and a good bonding situation, turned into the opposite. It taught me never to trust my superior. In addition, his escalation tarnished my reputation with the assistant chief, something that would be difficult to rebuild. He also made himself look like an incompetent leader to his own superior. Can you examine the scenario and see how he could have handled it differently?

Sir Isaac Newton's Third Law of Motion states that for every action, there is an equal and opposite reaction.[7] While he was applying that law in physics, the same principle applies here. When you decide it's time for you to give correction or discipline, make sure it's done for the right reasons, to help the individual. For every action you engage in will have a reaction, both immediate and long term.

When you do things the right way and for the right reason, then the reaction will be right, especially in the long term. Certainly, the opposite holds true as well. As far as self-discipline, I leave you with this quote from the first female champion of UFC Ronda Rousey, "The moment you stop viewing your opponent as a threat is the moment you leave yourself open to getting beat. You start thinking you don't have to train as hard. You cut corners. You get comfortable. You get caught."[8] You are your toughest opponent. Stay strong and never ever get comfortable.

PART 4

CUSTOMER SERVICE

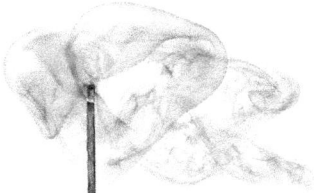

Chapter 20
ALWAYS TAKE CARE
OF THE CUSTOMER

Try to remember this your whole career: your job exists because of the customer! Without the customer, you don't have a job. Without the customer, there is no fire service. So, we owe all our thanks to them. They have actually given us the opportunity to have a job doing the most noble thing in life, serving humanity. This job also gives us an opportunity to care for those we love: a partner, children, family members, friends, and others. We are the blessed members of the fire service.

Service is what we are all about. Just about every single business has a customer service aspect to it. Interestingly, many times the fire department doesn't really see this side of their job so clearly, yet the fire service has the ability in their customer service interactions to truly change lives, families, and communities. We have the ability, with our service to others, to draw towns and cities together, even states and the nation as a whole.

Today, our country and the fire service itself face significant social issues in addition to the emergencies we handle. We may be called upon to handle fallout from protests and riots and other issues brought about by the unrest. But we are just as much a part of the community as everyone else, and we have a measure of authority and responsibility. We can use these to set a shining example, to help soothe the unrest and bring our communities together. It is not our job to cast judgment on anyone but instead to be an upright individual, a caring and loving human being and to give our community the best service they have ever experienced. We want those that we serve to feel we are all together, we are all part of the community, we are family and we are here to help.

Let me give you a few examples to ponder.

Holidays are always a calmer day at work. The city seems to be at such peace. Many people have the day off of work, so traffic is at a minimum and people are enjoying company and football in their homes. It feels like time was rolled back 50 years. Inside the station, the pressure is greatly reduced as well, the fire administration is off, the monthly requirements are suspended for the day, and after normal morning checkouts and duties, the men and women settle in and enjoy the peace.

Many family members come to visit, and throughout the station the sounds of young children can be heard as their moms and dads show them around. One Thanksgiving Day, the alarm suddenly went off. It was a fire. Upon arrival, we found a fire in the kitchen. The homeowner was roasting a turkey and grease spilled out of the pan it was being cooked in and ignited. The oven and the back wall of the kitchen were in flames. We put out the fire and opened up the wall to check for extension. In our community, we have natural gas lines running to each house, which is a rarity in Florida. We locked the gas lines, as is protocol.

When there is a fire, we are required to lock out the gas lines so the gas cannot be used again until the homeowner gets all the needed repairs done and the gas company inspects the repairs and ensures everything is fixed properly and operating safely. The gas company will then unlock the gas line and light the pilots. I explained all of this to the homeowner. She became visibly upset and kept saying, "No, no, no, I have to cook this meal, you don't understand I have to cook this meal!" I explained that she wouldn't be able to cook until she had the repairs done.

Her face seemed to crumble as she started crying. I tried to console her when she grabbed my arm, looked up at me with tears running down her worn face, shaking she said, "You don't understand, I have to cook this meal." She explained that she has not seen her family in over 20 years—they had stopped speaking to each other over some family issue. They had just recently begun to try to make amends and they were coming down from New York to have dinner that night with her since it was Thanksgiving Day.

I felt genuinely pained for her. I wanted to help her desperately, but what could I do? I thought about it for a moment before I had an idea.

"How much longer do you need to cook the turkey?" I asked.

"A couple of hours," she said.

"Why don't you give us the turkey and all the side dishes, and I'll cook them at the fire station. When they're done, I will call you and you can come pick them up."

She stared at me for a minute or two, completely silent as she seemed to be trying to gain composure.

"Is that okay?" I asked.

She was in tears and gave me a huge hug. We took her turkey and all the vegetables and cooked them for the next two hours. When it was complete, we called her, she came and picked it up and took it all home.

What do you think was one of the topics of conversation at her dinner table that night? *The Fire Department.* What do you think her thoughts will be about the fire service for the rest of her life? What do you think she feels about the community she lives in? Do you think she might give back to someone else? What example did we set for her?

I tried always to be cognizant of an opportunity to exceed someone's expectations from our service.

Hurricanes are something we are threatened with on a yearly basis in Florida. While we prepare for each threat out there in the ocean, relatively few hit us. One year, however, we got hit by a very slow-moving hurricane.

Although it wasn't very powerful, it lingered over us and caused heavy damage. After it passed, my crew and I went door to door in our zone to check on people's state of health and assess if they needed anything.

Many people were not at home, having left the city because of the impending winds. But for those who were, power was their biggest concern: the storm had disabled powerlines and the town was without electricity. There was no way to cool down in the hot Florida weather, which was dangerous for many. And people were worried about the food in their freezers going bad as well.

I determined that supplying ice would significantly help in both situations. I contacted the chief and persuaded him to get as much of it as possible. He had me come to the command center, where he, the police chief, town officials, and other officers were meeting, so that I could explain what I had seen and what I thought was a good solution. I was thanked and dismissed to go back to my duties while they discussed it. By the next day, my idea was adopted but modified. The town set up three strategically placed areas where any resident could get free ice. I was excited that the fire service had provided an above-and-beyond service for the community. Although I'd received negative feedback initially from some of the officers, many came to see the benefit and ended up appreciating it.

The biggest reward was seeing people's reactions. When we knocked on their doors to ask what they needed, they were shocked and grateful. Many residents of our community are older, and their faces, especially, lit up. They felt vulnerable and unsafe in their own homes as the power outage lingered, but the fire department was able to get them the help they needed in that moment. Can you imagine the sense of relief that brought? What do you suppose people's thoughts were toward their fire department from then on? What do you think they thought of the community they lived in? What example did we set?

Another year, after a hurricane hit and we lost power, I did some checks in town. The owners of a business on the third story of one building had prepared for the storm by having their windows covered with wood. Now they were without power *and* ventilation. When I knocked, they asked, almost jokingly, if there was a way to get the wood off the windows, as the contractor who put it up would not be available till the next week.

They looked surprised and hopeful when I told them I might be able to help. I called for the ladder truck to respond and upon its arrival, I asked the driver to put the bucket up to the third-floor windows and remove the wood shutters. We were used to doing that on our station windows, but everyone absolutely hated that job. The driver was furious. He didn't feel this was the appropriate use of the truck and he was upset that I'd called him out of the station. I couldn't understand how we could see things so differently. He gave me a hard time, but since I was in charge, he complied.

The people in the office were extremely grateful. Everyone thanked us over and over again. It was hard for that firefighter to remain grumpy with such a gift of thanks.

What do you think those office workers felt toward the fire department after that day? How did they view the community? Do you think they were more likely to pass on some goodness after what we did for them? What kind of example did we set?

But something even better grew out of that operation. That ladder truck driver was not my regular crew member, and he had a reputation for being grumpy and complaining a lot. Years later, I was stunned by a request of his. He'd had difficulties with several officers he worked under and had been bounced from one to another. The chief finally gave him an ultimatum. This was his last shot.

The driver requested to work with me. I was simply told, "You will be getting firefighter Higbee*" and had no say in the matter. I was saddened that I'd be losing one of my crew members to make room for him. I knew his personality would be a tough one to work with, but I am a believer that people want to be good at their jobs, they want to feel pride in what they do, and they want to help other people. At our first meeting together, one-on-one, he asked me, "Are you still doing that customer service stuff?"

At that moment, I thought he was already about to give me a hard time. However, I quickly realized he was moved by it and that he actually wanted to do more. He also loved to drill, which I had a reputation for as well. I believed that this individual was incredible raw talent and a very hard worker and that proved to be true. In fact, he proved to be one of the best firefighters we ever had in that department. He loved working with us and that was evident each day. He was not only very easy to work with, but he was also an incredible help. He has truly become a great asset for the department.

It was a joy to serve my city. When I got off work in the morning, I'd often visit the patients we had run on the day before to see how they were doing at home, or even at the hospital if they were still there. Other times I would visit them the next shift at their home to check up on them. If it was someone who didn't have a lot of help, I would visit them a month or two later as well. I did the same with customers of fire calls. I filled people's gas tanks with gas when they ran out on the road, I bought people groceries, and gave money to those in need.

I had so many wonderful experiences. So many. I cherish them now even as I write this. But the point isn't about me, it's to show you that the fire service needs to provide the best customer service it possibly can and to tell you that you cannot imagine what the response will be to your efforts. It will fill your heart for the rest of your life and you will be showing the right example to other firefighters. What if an entire department had the conviction to give the customer an experience far above their expectations? How amazing would that department be?

CAN YOU ADD VALUE TO PEOPLE'S LIVES?

I love reading and listening to Tony Robbins. I've learned so much from him, and I'll continue learning from him for the rest of my life. One of the things I like so much about him is his desire to truly affect those listening to him. If you ever get a chance to hear him speak, and I highly encourage you to do so, listen to how often he says, "I hope I can add some value to you." He will often start off a lesson that way and then finish it with, "I hope I was able to add value to you today," in one variation or another.

He wants people to walk away from their experience with him moved and exceeding the expectations they had going into it. That is the way we have to view our interactions with the public. Yes, they called 911, and we are here to mitigate their emergency, but can we add value to their lives as well? Can we exceed the expectations they had when they called 911? Can we leave them with the feeling that the fire department in their community genuinely cares for them, wants to help them, and will do whatever possible to make their life better? If we can, then they will do the same, giving back for you and for others in the community and country.

You know what happens when you toss a pebble into a lake—how it sends out ripples that keep going? Your actions are that pebble. When you add value to someone's life, when you significantly exceed their expectations, that ripple effect will keep on growing and keep on going. You never know what kind of amazing effects your service will have. So, keep doing things every day that shine as examples of what the fire service is all about. Make your brothers and sisters proud no matter where they serve in the country. Bring kindness to a fellow human being. Bring your community together, and hopefully the ripples of your service will never stop expanding.

WHAT CAN YOU DO?

1. First and foremost, recognize that service is a mindset. You must feel it deep inside: a desire to truly help humanity, to look past judgment or physical attributes and recognize that you are the person that can change a person's life for the better.

2. Speak with the chief of the department or other high officials about your desire and ask what can be done.

3. Always, on every call, look for ways to go above and beyond what

is expected, to leave people with a good feeling that blew away their expectations.

4. Take incredible pride in your job, your department, your city, and what you do for others.

5. Inspire others in the department to do so as well.

6. Remember it is the fire *service* you represent.

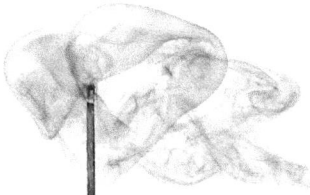

Chapter 21
IT'S ABOUT THEM

Among the most beautiful aspects of a fire department are the traditions. Traditions add wisdom, dignity, grace, and romance to the service. The hallowed halls of fire departments all over the country are filled with the history of the men and women who have walked the floors, fed the horses, checked the engines, cleaned the stations, played cards, and cared for the public. They are filled with memories of the men and women who gave the ultimate sacrifice in trying to save others, and men and women who've given the greatest gift by saving a life.

How fortunate we are to walk in the large and sacred shoes of those who came before us, who left us with the beauty, honor, respect, and tradition of a service we get to work in and live in every day. But while we are blessed to have the greatest job on Earth and centuries of tradition passed on to us, we have to be careful not to let that stagnate us, for this could serve as a drastic negative for the department.

To this point, I've highlighted a few lessons we can learn from the business world. Why? Because the fire service is a business. Sometimes we don't understand that. Yes, we see it as serving the community and sometimes we act as if we are an autonomous agency. However, to city managers, the fire service is most certainly a business and so it is to many city residents who understand that they are actually in control.

Society is changing. If you've been around for a while, like me, you remember a very different business world. When you ordered something, you were told to "allow six to eight weeks for delivery." If you weren't satisfied with the service or product you purchased, you were told to take your business elsewhere. Customer service was not even an afterthought, it plainly did not exist. When I was growing up, if you ordered a burger from McDonalds, you took it the way they made it. If you didn't like pickles or ketchup, too bad—that was the way it came. If you don't like it, don't eat there. In fact, one of the ways that Burger King was able to grow its business against the already established McDonalds was to allow customers to order what they wanted, "Have it your way at Burger King," was their slogan. How things have changed. Now things are delivered to you within a day or two and for free. People rate everything on the internet, and if service isn't fantastic and the negatives build up, it can be an end of a company.

Bad service, a single employee's mistake, or negative social interaction can go viral and crush a company or cost them millions. Customer satisfaction has become king. Businesses that have not kept up, that were slow to respond to what customers wanted, have disappeared. In the 1990s, Nokia was the king of the cell phone business. By 2007, it controlled 40% of the market share, something no one else has achieved even to this day.[1] I was extremely active studying the stock market back then. Everyone was talking about the day when there would be a phone with a touch screen and wondering why it was taking so long for it to happen. Nokia apparently developed such a phone but didn't think people would buy it.

The cell phone industry was destroyed by a computer company that decided to get into the cell phone business. That computer company, Apple, would design a phone that customers wanted, a touch-screen phone that would become the standard. Does anyone have a Nokia phone anymore? Apparently, so few do that they don't even show up on market share reports.[2]

In the mid-1970s, for the first time ever, it became possible to rent a movie on video and watch it in your home. Suddenly, video rental stores were popping up all over, but they usually stocked just a few copies of popular films that typically were already rented out when you tried to get one.[3] Sometimes it took months to see the one you wanted. Then in 1985, along came a company called Blockbuster with a winning business strategy: give the customer what they wanted.[4] They stocked 25, 50, 100 or more copies of each of the newest movies.

Blockbuster's strategy would dominate the video rental business so much that they put almost everyone else out of business. By 2004, they would have 9,000 stores worldwide.[5] But along with their incredible success they quickly forgot what had made them successful: the importance of satisfying the customer. Blockbuster became known not as the king of video rentals, but the king of late fees.[6] To many, it appeared they designed their business precisely to get these fees.

It was, in fact, these very late fees that so irritated a man named Reed Hastings that he started a small movie-rental mail service called Netflix. Their hook? No late fees. They began to gain traction and even tried to sell themselves to Blockbuster for $50 million.[7] Blockbuster balked at the deal, after all, in their mind, they were the king.

Within several years, Netflix had begun to take market share from Blockbuster. And Blockbuster actually tried to emulate Netflix with a mail service, but it was too late. They had irritated customers too much with their late fee program.[8] Today, Netflix is a juggernaut worldwide, currently valued at $296 billion[4] and many young people have probably never even heard of Blockbuster. The business world is littered with plenty of additional examples.

But we are the fire service you may say. Yes, but the fire service is a business to those who pay our salaries. How do our customers feel about us? I remember reading articles online about home fires that fire departments all over South Florida had battled. I was shocked when I scrolled down and read people's comments. Certainly, there were positive ones, but the majority were about how the firefighters were always in the local grocery store, or that they sit around at the firehouse and never work, or they're always on their boats—on and on. I couldn't believe it.

Have you read the comments posted about the fire service in your town? Does it mean that people in our communities don't like their firefighters? No. It's human nature that people are more likely to speak up when they're upset about something than they do when they are happy, and some are just jealous or complainers. But some comments are legitimate complaints, so we cannot discount them or dismiss them.

If we don't focus on what will satisfy the customer and make them happy, we may find ourselves like those businesses we highlighted earlier. This is especially true in economically difficult times. After the economic crises in 2008, our city gutted our benefits. They even met with a private fire department about taking over our service. If that had happened, we would all have been terminated. Thankfully, it didn't happen. And who would ever have imagined that residents and city managers would even consider defunding police departments? I would never in my life have imagined that and I am not in favor of it, but whether you agree or not, the point is, customers are rightfully flexing their muscles.

It's their money. The department belongs to them and we want to make sure we bring top service to them, service far beyond their expectations. We want to make sure we are satisfying their wants and needs, otherwise we will become their focus one day.

We have been blessed in so many ways. We have one of the best jobs in the world. We get to serve people every day. We get to make a tremendous impact upon their lives. We have inherited the hallowed traditions and dignity of those who walked before us. But never forget that we are here because the customers need us and pay us to be here. It is about them. As a leader you can always set the example. How can we proceed? Proceed to the next chapter.

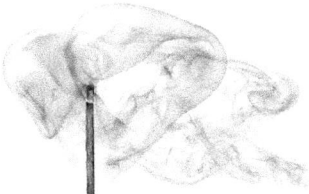

Chapter 22
LOVE YOUR JOB - SHOW PRIDE

It was a long, exhausting day training in bunker gear in the Florida sun with a heat index of 100. We also ran emergency calls back and forth all day. It didn't stop. On a normal day at the department, I didn't get to bed until about 0200 (2:00 a.m.), but that night I was completely exhausted. I ate, ran another call, took a shower, and finally at about 2200 hours I hit my bed. But the alarm was not going to give us a break that night.

At 2330, the next call came in. At 0200, the alarm rang again. Down the pole, into the engine, and away we go. We were on our way to a dance club. Dispatch informed us that the patient was a female, early 20s, passed out. When we arrived, most of the people had already cleared out. We entered the club, saw the patient slouched in a chair, and quickly walked up to her. Her inebriated friend interrupted our path and drunkenly slurred out that her friend was all right, that she just drank too much because her boyfriend broke up with her and that the "club is a a$$hole."

Yes, she said it that way. The lieutenant started to ask the patient questions, but she only moaned a response. He told us to help her up and put her on the stretcher. As my friend and I each grabbed an arm to lift her, she turned and spat in my face and yelled, "F@ck you." I held on to her until

I was relieved by a police officer who quickly took over the questioning. I went to wash my face.

I was utterly disgusted. There I was trying to help her in the middle of the night, exhausted beyond belief, and she spat into my eyes, nose, and mouth, putting me and my family in danger. But that was my job. Fires are not the only risk to life and health we face. I've been stuck with a used IV needle, splattered with bodily fluids countless times, vomited on multiple times, and almost hit by cars more times than I wish to think about.

This patient wasn't the only one who caused us stress. I was part of a crew that was wrongfully sued for "stealing a watch" from a man who we brought back to life. I have been part of crews that were falsely accused of stealing a briefcase, a telephone, and even sued for the death of a hedge that was killed by the heat of an $11-million house fire my crew put out. An older Navy man wanted to box me once, a younger man wanted to karate fight me. I've been cursed out multiple times, screamed at and threatened more times than I can count—and so much more.

But I LOVED MY JOB! Most of those people were either drunk or on drugs, the others, well, perhaps they were just having a bad day. As I write about these incidents and relive them with my wife, we crack up. What a job. Unfortunately, you see people at their worst sometimes, but I would have to say that these were by far the minority of incidents with people I encountered. I prided myself on doing my best to give my customers, my patients, the kind of care that exceeded their best expectations.

While I received hundreds of thank you letters throughout my career from many of those I served, the greatest thanks I had was knowing that I made a big difference in a person's life. There is nothing like it. We are, in my opinion, the most privileged people in the world, precisely because we do see people at their worst and receive the training and incredible tools to make a difference, a positive impact on their lives. It would be nice if everyone appreciated our efforts, but those who don't shouldn't change how we feel about what we do: their bad moments give us an opportunity to change their life around.

AN OPPORTUNITY OF A LIFETIME

The call came in early afternoon, "a possible drowning" on the north side of the island. I am never comfortable when I get these calls even though most of the time it ends up being a false call: either the person got out of the water by the time we got there, or someone mistakenly thought they saw someone drowning.

We arrived and dragged our equipment down the beach. There was a young man lying lifeless, surrounded by friends who were crying and anxious. The young man had been surfing and went under and didn't get back up in time. His friends dragged him out of the water and onto the sand and called 911. It was on the farther end of the island, so there was a good amount of time between the time he died and the time we arrived.

We immediately began CPR, hooking up our EKG monitor and other equipment. We intubated him and got IVs established. We pushed drugs to help revive his heart as we continued to perform CPR. I called for a helicopter for quicker transport. I contacted an engine from another zone and had them respond to set up a landing zone at the golf course on the north end. We would meet the helicopter there. We carried the young man on a backboard, dragging all the equipment up the beach until we got to the road where we loaded him onto a stretcher and headed for the rescue. By this time, we were able to get a heart rhythm. We continued to use the EKG monitor to supplement his heartbeat and bagged him (forcing oxygen into his lungs) as we climbed into the rescue and headed for the landing zone.

As we pulled into the golf course, the helicopter was waiting for us. It was a fast and successful transition and the pilot was up and in the air. This allowed the patient to get to the hospital in less than 10 minutes—it would have taken us 20 to 30. It doesn't sound like a huge difference, but when it's a matter of life or death, every single minute is critical. Although we'd brought him back to life, his situation remained critical and he could have gone the other way.

Three days later, he walked out of the hospital. He was 18 years old.

His whole life was ahead of him. He had a family that loved him and was watching him grow into manhood. We brought that young man back to life and in very difficult circumstances. It is something I will never forget. He would not be the only one we revived, but he was the youngest, the one with the most life ahead of him. His bad day gave me the opportunity of a lifetime.

EXCEPTIONAL SERVICE

Henry Ford wrote, "As we serve our jobs, we serve the world."[1] Those words ring very true in the fire service. Your actions affect so many people: your patient, all those who love that individual, and everyone those people interact with in the future. If an event should make the news, which many of the emergencies we handle do, then an impact will also be made on all the people that read about it or hear about it on TV or hear others talk about it. They will have an opinion on what happened and the way it was handled. Your actions have the ability to touch and impact so many.

We will talk about loving what you do and doing it with pride in chapter 27, but there we'll focus on you personally and the fire service. In this chapter, we're focused on customer service. If you genuinely love what you do and you take pride in doing it, all your interactions with those you serve will reflect it, even when no one sees it.

A couple of days ago, a video popped up on my YouTube homepage.[2] Someone had posted footage from their front-door camera of two delivery people from two different companies in the same day. The first courier came close to the door and dropped the boxes to the ground. The second courier was from the USPS. He brought his package closer to the front door, then maneuvered something in front of it in order to block its being viewed from the street. He then picked up the boxes dropped off by the other company—which were now getting wet from rain—and placed them closer to the door as well.

He probably had no idea he was being videoed. And he probably didn't care. He cared about what he was doing, he cared about his customer. He wanted to make sure the packages were close for the customer to get, that they were hidden from the road so no one would steal them, and that they weren't in the rain getting wet. He didn't care that he was helping his competition, he cared about the customer and their experience. That man is proud of what he does for his customers. He doesn't need them to see it or even express appreciation for his efforts, he loves what he is doing for others. It is obvious.

From time to time, a customer will see *your* love and pride in what you do, and when they do, imagine what they will think about their fire department and the fire service in general.

Have you ever gone to a restaurant and the waiter or waitress was just exceptional? They were extremely attentive to your needs, they were warm and friendly, they explained everything to you and made you feel really welcome? Didn't you enjoy the experience so much more? Have you ever gone to purchase something when you were unsure about exactly what you needed but the person helping you was extremely knowledgeable, took their time, helped you find what you needed, and even explained some of the product to you? How much more at ease did that put you? Those individuals loved what they did, they took pride in doing an exceptional job for you and you benefited greatly from that.

We can do the same for those we serve. When you love what you do for others and you serve them with pride, you will look for opportunities to care for them, to go out of your way to help them, and to do the very best possible job—better than they expected.

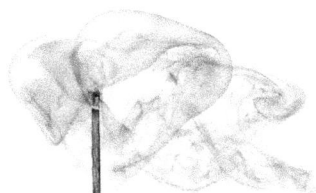

Chapter 23
NEVER LIE TO THOSE YOU SERVE

We discussed the impact lying has on you, your character, and your ability to lead in chapter 7. But what about your customers, your patients, and those you serve? How open and honest should you be with them?

We should always try to be as open and honest as possible with those we serve. When responding to an emergency, people are already tense, scared, worried, nervous, and desperate for help. If you are dishonest with them when they confront you, and they see that, it will only add to their stress. And by not being truthful, you take away from the honor and dignity of the fire service.

AN HONORABLE PROFESSION

We responded to a call of a patient not feeling well. We arrived at a massive house on the beach—not an unusual house in our town. But with such houses it takes a few minutes to get in and get out because of their size and security details. We were led to the back of the house overlooking a picturesque ocean scene. Walking up to the gentleman, he stated he was weak and not feeling well.

After we quickly stabilized him, we brought him to the back of the rescue vehicle, where we checked a few more vitals before driving him to the hospital. We felt rushed because of the delay in getting him back to the rescue: he was a high-profile U.S. senator, so there were a lot of aides and security people around him, and these too caused delays. As we got him into the back of the rescue, I went up front to drive. One of his aides ran up to the window and handed me something for the senator.

As she walked back around the front of the rescue, I turned and reached behind me to give the officer the item for the senator. When I did that, I accidently stepped on the airhorn. (There was a button on the floor to activate it. This would be removed in all future rescues as it was a terrible design flaw.) I heard it, but it took me about 10 seconds to realize that I was stepping on the button. As I took my foot off the horn and looked up, there was the aide holding her ear in front of the rescue, right where the horn is its loudest.

I immediately became super nervous. She walked back to my window and said, "You just blasted out my eardrum." I was so nervous and young, and due to the combination of working a high-profile patient, the time delay, and what I had just done, I couldn't think straight in that moment. Instead of apologizing and explaining what happened, I said things that didn't make me sound very empathetic about what she had just experienced. She looked at me angrily and walked away. I was already nervous from our delayed time on the scene, but after I made that incredibly dumb statement, I became overwhelmed with anxiety.

During the ride to the hospital and back all I could think about was what just happened. This was a senator's aide. I knew for sure I would be questioned and then disciplined and perhaps even fired. What should I say, I thought to myself? Should I say she misunderstood what I said? After all the unit is quite loud to begin with and she had just had her eardrum blasted, perhaps she misunderstood what I said.

I was terrified, but I resolved to tell the truth and lean on the fact that

I was still relatively new, I was young, nervous already, and it was just a mistake. But that questioning never came. The aide must not have said anything. (If you are that aide and reading this book, I am sorry and thank you.) It was such a relief that nothing came from that mistake. However, what felt better was that I resolved not to lie. I was proud of myself. Even though I had made a mistake and then made it worse with my big mouth, if I was questioned, and didn't lie, they would have known that I value the fire service as an honorable profession.

If you are an officer already, and those you lead see you being dishonest with our customers, it sets an example for them that it is okay to lie to those we serve, and it also diminishes your leadership in their eyes. It also diminishes their respect for that person. In essence you teach them that our patients, our customers don't have to be treated with the utmost respect and honor. We send a message to those we lead with all our words and actions, and we are responsible for training the next generation of leaders. Is this the example you want to set for them on how to treat those we care for?

If you get caught lying, it can also have other significant repercussions. First, it could mean your job, but that should never be the motivation not to lie. You should always want to be viewed with the utmost moral character. Second, it will leave a mark on the whole department and all your brothers and sisters in the service. Certainly, the whole service can get marred by a few bad apples.

That does not mean we will always be able to share any and all information we have with those we serve. Many times, there may be reporters at scenes where information should be disseminated only through the department personnel assigned to do so. Neighbors might ask about the condition of your patient, but that is not information you can share—both for the dignity of the patient and because of Federal HIPAA laws.

Many times, you see people at their very worst. It would be cruel and unethical to disclose that information. Some have been tempted to take pictures while on emergency scenes and post them to their social media accounts. Depending upon what a picture shows, this might be illegal, against department rules, unethical, or just downright mean. We love serving our customers and making emergencies better for them. We would never want to do anything to hurt someone. A first responder in an agency close to where I live was caught taking severed body parts from a scene, intending to keep them. This is completely unacceptable. Nothing that belongs to the patient is ours for the taking.

Today, people have an almost insatiable thirst for any information on celebrities and famous people. Throughout my career, our city has routinely had visits from such: sitting presidents, ex-presidents, politicians, actors, actresses, athletes, singers, TV personalities, high-profile business women and men, and every other kind of famous person you can imagine. Many of our city residents are high profile people, as well, and inevitably some of them have emergencies. Our city has seen incidents that were all over the news for many months. You see people in vulnerable positions and learn things about them that they want kept private. It is imperative that you respect that.

Even now, years out of the department, I get asked about high-profile cases and people all the time. I have never divulged a word and never will. To me, the residents I served placed me in a position of honor to help them and I will never dishonor that, I will never lie to them, and I will always protect their dignity.

Perhaps you are familiar with the Hippocratic Oath—an oath of ethics for physicians that was established in ancient Greece, about 2500 years ago.[1] It required new physicians to swear to uphold specific ethical standards. It has been rewritten through the years to adapt to specific cultures using it and there is a modern version for today's world.[2] It is not a document that is legally binding—it's more like an ethical signpost—but it is beautiful. It is women and men promising themselves that they will use all they have learned in school and all they will learn through the course of their career to treat those that they care for to the best of their abilities and with honor and respect and dignity. I think it would be appropriate and dignifying if we as the nation's firefighters created and adopted our own national oath where we make that same promise. Perhaps we can start a nationwide conversation on this.

We firefighters have the honor to care for our great nation and its incredible people. We are a beacon, a shining example, we are held to a higher standard, and we should proudly hold that always. Being dishonest with those we serve destroys that shining example. Even if we are unable to make a nationwide oath, perhaps it can be something you do within your department or with your crew. It could bring people together as one, instilling in them a sense of pride for their customer service.

No matter what, always hold yourself in a dignified way, knowing you never need to lie, it is beneath who you are. Always protect the dignity of those you serve.

PART 5

LEAD THE DEPARTMENT

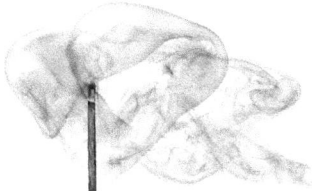

Chapter 24
BE HUMBLE

Humble? Yes. Many will tell you humility is the last thing you should have, that being humble is a sign of weakness and a weak leader is an ineffective leader. Is that true? Absolutely not. Do not confuse humility with weakness. The fact is, it takes greater strength to be humble and to act with humility then it does to be proud. Humility is sorely lacking in our society today, and that is precisely because it takes strength and character to act in such a way. Let's talk about a few points: Why is humility so important? How do you achieve it? The promotion problem. And, is it possible to be humble and still have pride in what you do?

WHY IS HUMILITY SO IMPORTANT?

The simple answer to this question is "because our fundamental job is to serve others." If you think it's not, you have missed the purpose of the fire service and what makes it great!

You and I are servants to those we work for, the public, our bosses, and our firefighter brothers and sisters. Does the word "servant" make you uncomfortable? It shouldn't, it should make you proud. Serving others is neither demeaning nor belittling. The president of the United States *serves* the people of America. Would you say that is a demeaning position? Obviously not. And this is where humility comes in. *It takes genuine humility to see yourself as serving others, to desire to do so, and to do it the right way.* Let's examine a few examples:

In his "The Drum Major Instinct" speech, Martin Luther King Jr. spoke about what he wanted said about him at his funeral. With all that he was doing and accomplishing, he said, "I'd like somebody to mention that day that Martin Luther King, Jr. tried to give his life serving others."[1]

He wanted to be remembered for serving others; that's humility. Showing his humility further, notice he used the word "tried" to give his life instead of "gave" his life. Certainly no one else would use any word other than truly "*gave* his life serving others" when speaking about Martin Luther King Jr., not just in death but in life a well. He continued on in the speech to say, "I want you to say I tried to love and serve humanity."[2]

Serving others is not only one of the most noble things we can do with our lives, it is also the *whole reason for leadership*. We are given opportunities as firefighters—and especially as officers—to serve humanity every single day.

Do others use the word "humble" to describe you? Navy SEAL Chris Kyle, "the most prolific American sniper of all time" said about those who lead him: "I have had good officers. But all the great ones were humble."[3] Notice he said *all* the great ones were humble, not some, not most, but all. In other words, in order to be a great officer, in his opinion, it was essential to be humble. Our goal is to be a great officer. Could you lead a Navy SEAL like Chris Kyle into battle? And if so, would he say you were a great leader? Well, humility would be a quality you would need to possess for that to be so.

In speaking about Major Dick Winters, Commander, Easy Company, 506th Parachute Infantry Regiment, another great leader, Captain John Kugler of the U.S. Army stated, "It started with him doing the right thing and it spread, but he was a humble man. Heroes are humble."[4] Simon Sinek referenced Brigadier General Lori Robinson, saying she "has shown me what the humility of great leadership looks like."[5] Jocko Wilkins, retired Lieutenant Commander Navy SEAL team 3, repeatedly speaks about the importance of humility for a leader in his book, *Extreme Ownership*.[6]

The Walt Disney Company employs over 200,000 people.[7] Imagine leading a company with that many people. Former Disney CEO, Bob Iger stated, "The first rule is not to fake anything. You have to be humble, and you can't pretend to be someone you are not or to know something you don't. You are also in a position of leadership, though, so you can't let humility prevent you from leading. It's a fine line, and something I preach today. You have to ask the questions you need to ask, admit without apology what you don't understand, and do the work to learn what you need to learn as quickly as you can. There is nothing less confidence inspiring than a person faking a knowledge they don't possess. True authority and true leadership come from knowing who you are and not pretending to be anything else."[8]

Along with reiterating the necessity to be humble, Iger defines another reason for its importance. He emphasizes "not to fake what you don't know but instead learn it."[9] From the time when we first enter a new job and through our continual changing circumstances as life goes on, there will always be things we don't know. In other words, there will always be things we need to learn (this is especially true if we want to grow). Humility makes it possible for you to learn from others, it is essential, and it encourages learning. Ralph Waldo Emerson wrote, "Every man I meet is my master in some point, and in that I learn from him."[7] It is virtually impossible to live that line without humility.

HOW DO YOU ACHIEVE HUMILITY?

So, how do you practice humility as a firefighter or an officer? Well, it starts with a state of mind. Humility is not something you can put on, something you can show from time to time. It is something you have to work on becoming internally, a personality characteristic, an intrinsic value. So, the first step is *evaluate yourself.* Ask yourself, "Am I a humble person?" Ask those around you. Ask your family or a close friend who will give you an honest answer.

Do you see the good qualities in others, even those who may have mistreated you in the past? Do you view others as superior to you in certain aspects? Can you submit to the direction of those superior to you who may be younger, much less experienced, or are a different gender or race without making disparaging remarks about them outside of their presence? Are you able to take criticism from subordinates and honestly evaluate it—without taking offense or getting defensive—and work on changing those things about you?

Can you admit when you are wrong without it being a big deal? Do you admit when you make a mistake and accept the consequences without pointing out other people's faults? Do you not judge others based on their nationality, ethnic background, skin color, sexual orientation, gender, religious preference, or financial station in life?

In a podcast interview, Jamie Dimon, CEO of perhaps one of the greatest banks in history, JPMorgan Chase, said this when discussing leaders who lack humility: "At the heart of (arrogance) is some insecurity."[10] Ultimately, being humble or not is 100% about you. You must search your inner feelings and you must work on what you find as a problem. The questions above are a great start.

Henry Ford stated in his autobiography, "None of our men are 'experts.' We have most unfortunately found it necessary to get rid of a man as soon as he thinks himself an expert— because no one ever considers himself an expert if he really knows his job. A man who knows his job sees so much more to be done then he has done, that he is always pressing forward and never gives up an instant of thought to how good and how efficient he is. Thinking always ahead, thinking of always trying to do more, brings a state of mind in which nothing is impossible. The moment one gets into the "expert" state of mind a great number of things become impossible."[11] When we recognize we always have room for growth and that there is always something we can be taught by each person, it will help us to become and remain humble.

THE PROMOTION PROBLEM

Promotion into a leadership position brings with it two unique problems. First, is the desire to show everyone you know what you are doing, that there is a reason you were promoted. The second is buying into the hype that you are special. Both stem from a lack of humility. Both reflect poor leadership and neither inspires others to greatness.

1. When someone is promoted, sometimes they immediately want to show everyone that they know what they are doing. They put on airs that they don't make mistakes yet are terrified on the inside that they will make a mistake. Carl Jung addressed something similar when he wrote, "There appears to be a conscience in mankind which severely punishes the man who does not somehow and at some point, at whatever cost to his pride, cease to defend and assert himself, and instead confess himself fallible and human. Until he can do this, an impenetrable wall shuts him out from the living experience of feeling himself a man among men."[12]

   When you are promoted, it's a whole new job. Learning how to lead takes time. In addition, learning all the aspects of your new job takes time. When you pretend, others see it. People are not stupid. All your pretension does is scream your insecurities and your dishonesty. That works against your becoming a great leader.

   Admitting that you don't know everything and even relying on others for their specialties shows your humility, your reasonableness, and your ability to recognize and utilize others for their efforts. This will draw your team to you. In addition, your bosses already know you don't know everything and that you're going to make mistakes. They were in the same position themselves. They felt the same way. Be yourself. Work hard to learn and be great at your new position. Don't be dishonest or insecure when taking the lead. As Dr. Jung stated, then you will know what it feels like to be a man or woman among your peers. You will indeed like who you are, and others will like you too.

2. Buying into the hype. This is a phrase I have used through the years to describe what I have seen when people began to think too much of themselves or their abilities. When individuals are promoted into a position of authority, they sometimes begin to think it makes them better than others or better than those they lead. Nothing leads to a crash in a person's life faster or deeper.

Addressing the issue of promotion, in the podcast I mentioned earlier with Mr. Dimon, he quotes John L. Weinberg of Goldman Sachs, who said, "Some grow into it and some swell into it."[13] Perhaps you may have seen individuals "swell" in the job until they thought they were bigger than the position. Again, a lack of humility will feed this and as it grows so does your ineffectiveness to lead others and to inspire them.

We are all the same, we are all equal. Never start thinking you are better than another person. The minute you start that, you begin to be a detriment to those you lead, the department, the people you serve and yourself and your crash is on its way.

IS IT POSSIBLE TO BE HUMBLE AND STILL TAKE PRIDE IN WHAT YOU DO?

Most definitely. You do this by giving it your all, never second best. You want to do the best you can, but never at other people's expense or by being deceitful. Can you give a project 100% of your effort, or as close to that as possible, and then handle some critique of it or criticism? That is having pride in what you do while being humble.

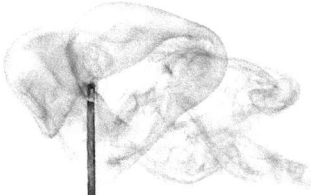

Chapter 25
COURAGE

"The ultimate measure of a man is not where he stands in moments of convenience, but where he stands in moments of challenge, moments of great crisis and controversy." -
— Martin Luther King, Jr.[1]

One word can sum up what Dr. Martin Luther King is speaking about: courage. It is defined as the ability to do something that frightens you. Also, it is strength in the face of pain or grief. [2] When speaking about firefighters, the word "courage" often comes up. How many people will run into a building that is ablaze when everyone else is running out? That is just one area where the call of duty demands courageous hearts to act with quick decisiveness in the face of death.

Besides courage in the face of emergencies, it takes courage to be a true, effective, and good leader. Even in the daily minutia of fire service life, it takes courage to do the little things well throughout the day.

"Courage is not limited to the battlefield or the Indianapolis 500 or bravely catching a thief in your house. The real tests of courage are much quieter. They are the inner tests, like remaining faithful when nobody's looking, like enduring pain when the room is empty, like standing alone when you're misunderstood," said Chuck Swindoll, a Christian preacher.[3]

It takes courage to always try to do the right thing, especially when it is expedient not too. It takes courage to examine oneself and work on the areas of deficit, especially when others see it. It takes courage to admit to others when you are wrong, especially when you are leading. It takes courage to care for others, especially when they don't treat you the same. It takes courage not to speak negatively about the department or others, especially when the rest of the group is doing just that. It takes courage to serve others, especially when they look down on you. It takes courage to be a good follower, especially when others mock you for it. It takes immense courage to stand out on your own when it's necessary, especially when you feel it will be used against you in your career.

Courage is needed throughout our daily activities and maybe even times we least expect it. Why? Winston Churchill said, "Courage is rightly esteemed the first of human qualities...because it is the quality that guarantees all others."[4] All of our actions are affected by our courage. All of the areas of leadership and character that we discuss in this book take courage to build on as a foundation.

If we wish to change ourselves, or work on individual areas of need, trying to do so without courage can be futile and even push us back further. In addition, others are watching us. They know when it takes courage to do something, and when you fail, they see it. That is especially true of your subordinates. If they see that you lack courage in smaller areas, what do you think they believe your reactions will be in bigger areas, including emergencies? The opposite is true as well: when those you lead see you in a position that requires true courage and you act accordingly, it builds their respect for you and deeper trust. Acting with courage not only sets an example, but also builds trust. In fact, trust cannot be built without it.

But what do you do if you are genuinely scared? What if you lack courage? How do you build courage? The answer is the same for all three questions: reach out and do it anyway. Do the right thing. *Courage only comes into play when you are afraid. It doesn't take courage to do something you are not afraid of.* So, when you are presented with a situation and you are afraid, you are nervous, understand that if you press forward and do the right thing you will build courage. It will become easier the next time you are presented with a situation that scares you.

Each time you are confronted and push through, you build courage. Do you remember, when you were very young, how scared you were on your first day of school? If your parents had allowed it, you would have stayed home permanently—but you went and each day got a little easier. How about the first time you drove a car? My first experience driving was when I took a driver's ed class after school. I was 16. On the first day of class, we were already going to do some driving. A friend, myself, and three other kids from school all piled in the instructor's car and took off. My friend went first. I was next. I slipped into the front seat. I had never driven an inch before in my life. He told me to go and then led me onto the Belt Parkway in Brooklyn, New York. I couldn't believe it, it's one of the busiest highways in New York City. Was he sadistic?

Every single muscle in me tensed like I had never felt before. I don't know how I didn't crush the steering wheel I was gripping it so tightly. But it only took a few lessons before I was much less nervous, and it finally became second nature. I was never nervous again.

Do you remember when you first started in the fire service? You reported to the department, you didn't know a soul. Everyone seemed to know each other well and know what to do each day. Perhaps you felt there were a million eyes watching every step you made. You were uncomfortable, nervous, scared, you did not want to screw up. But each day that went by got easier. As you faced each challenge you became less afraid and more comfortable.

Sometimes we are just scared of the unknown. Courage takes us through it and we then realize there is nothing to be afraid of and we are better and stronger when we get to the other side. Courage is the choice and willingness to confront agony, pain, danger, uncertainty or intimidation.

When you face something with courage, when you do what is necessary even though you are scared, it builds your character. It builds your self-esteem. It gives you a better understanding of yourself and life. It is a building block to greatness.

You can reach out to a mentor for help when you feel scared about a situation. Talking it over with a trusted mentor can help you see the situation the right way and be the push that gives you the courage to face the challenge successfully.

Another thing to do is to look at past accomplishments, past successes, past incidents where courage carried you through. Let them help push you through the next one as well. And, finally, remember the greats. Where would we be without the great accomplishments of Martin Luther King Jr, Abraham Lincoln, George Washington, Harriet Tubman, and many more. They accomplished all they did on a foundation of courage. They are the same as you. You can do this. I know you can and so do you.

Do not go gentle into that good night!

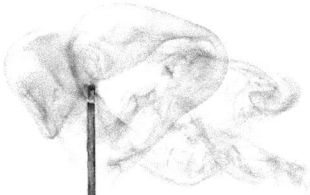

Chapter 26
DETRACTORS

Along the way, you will encounter people who will create obstacles for you when you try to better yourself. When you begin to reach out or better yourself some people won't like it. People might judge you and talk negatively about you. Sadly, some people will be jealous of you making changes in your life, becoming a better firefighter or officer or a better person. People very close to you might even discourage you from your progress, thinking they are helping you. Of course, not everyone will feel this way. Your true friends will be happy for you and you may even have a positive influence upon them. But be prepared for those who might work against you.

They may start rumors behind your back, make things up, exaggerate negative things that have happened in your career, or even straight up bad mouth you to others and superiors. They may act one way to your face while behind your back it's a whole different story. Some will secretly wish for you to trip up, and if you do, they will gladly spread the news to as many people as they can. Don't think this is exclusive to the fire service, this same dynamic breeds in corporate America, public employment, small businesses, and even in people's personal relationships with friends. This kind of behavior can cut deep. It can even derail you if you let it.

Why is this the case? It's human nature, or as author and speaker Julie Broad puts it: "If you want to have an impact upon others, you must stand out. But when you stand out you will be judged, and that can be terrifying."[1] And it can be hurtful. As I mentioned earlier, I was the pension representative for our fire department for about a decade. When I was elected to the pension board, it was a third-rate pension.

I knew a little about investing in stocks and the stock market but nothing about how pension systems work. I worked tirelessly for years to get a good understanding of how the system works, what the vendors do for us, how we invest, and what benefits we could negotiate with the city. I paid for and went to the UCF Certified Financial Planning program and countless other courses to increase my financial knowledge. I oversaw the management of the pension money and negotiated retirement benefits for the police officers as well, along with a police representative. When I started, our pension calculation was 2.25% per year of service. That means if a firefighter worked for 25 years, they received 56.25% of their pay. Before that, it was 2% a year. So, most of the firefighters had some years at 2% and some at 2.25%—they would not get 56.25% in the end. It would be less.

I was not paid to be the pension representative, but I worked well over 15 hours a week on it for years. I negotiated a change to the accrual, first 3.25% and then 3.50%. But what made it even nicer is that I negotiated the benefit to be retroactive. That meant that a firefighter who had some years at 2% and some years at 2.25% now had all their years at 3.5%. People's pensions drastically changed because of my time, effort, and work. And that was just one huge impact that I had.

Every two years, there was an election for the position. As long as I wanted it, no one ran against me, I took that as evidence that I was doing exceptional work. However, I didn't always get all the votes. One year someone wrote in: "Anyone but John Cuomo." Now, what did I care? I ran unopposed and I received every single vote except two that year. But it still hurt my feelings. This individual apparently did not like something about me but he or she was too cowardly to run against me, too cowardly to write their own name in, and full of hate to the point where they wanted to let everyone know, anonymously, that someone didn't like me. They did not have to vote at all if they didn't want to, but instead they chose to take a swing at me anonymously for all to see. Sadly, this would not be the only time a person took to anonymous writing to hurt someone in our department.

That highlights a second reason people may speak badly about you. As I pointed out earlier in the book, I didn't start off my career on a stellar note. I had a strong personality and I was naïve. I was not humble. I ripped back into people who ripped into me. I lacked proper respect for tradition and the older firefighters. I made mistakes, learned painful lessons, and changed tremendously as I grew up. I always apologized to people who I offended, and I worked on my respect and humility, however, that doesn't change the fact that along the way I hurt people's feelings, I offended people. Some of them may not have gotten over that, especially when they saw me advancing, while they, perhaps, were not advancing themselves.

When judgments come, unfair criticisms, rumors, slander, passive-aggressive comments, and discouragement, what do you do? You absolutely must not let these distractions stop you, slow you down, or affect your progress.

First, know that these things more than likely are going to happen. Being ready for them will help you to not get derailed. The people who mock or judge or criticize or talk negatively about you are not the ones chasing or achieving greatness, creating success, and having a lasting positive affect on the department and others. Keep your goal always in front of you: you want to be a great leader and have a significant positive influence upon the department.

Second, you cannot allow criticism to slow your progress. It will hurt, everyone wants to be liked—even by people who you don't really have much in common with. No one wants to hear that people are saying nasty things about them. Even those you may help or have helped may turn on you. You cannot take it personally. Keep your eyes focused on your goals and moving forward.

Third, if you hear of things, try to fix them if you can. You don't want things spread about you that are untrue. People may start to believe them.

Fourth, never ever retaliate. Never. Engaging in insults back and forth is like staying on a sinking ship. It is useless, it isn't going anywhere but down, and it will take you down with it. Stay away from it, or if it should start, rectify it and move forward.

Fifth, self-examine. Did I do what they are saying I did? If so, fix it when you are wrong. If it is just slander or negativity or disrespect, try not to be affected by it. When Jesus visited his hometown of Nazareth, He was treated disrespectfully. He stated, "A prophet is not without honor except in his hometown and in his household."[2] Jesus healed tens of thousands of people, raised the dead, was kind and humble, never sinned against anyone, freely fed thousands of people, and yet he was treated badly by people who knew him. It happened to him, so who are we? I am nothing compared to him. It will happen, know you are in good company.

Sixth, set the example. You are the leader or reaching out to be one. Help others to reach out, to progress. Encourage them. Never say negative or false things about others. Celebrate when someone in your department wants to take the department to the next level of greatness. Do what you can do to be a help.

We spoke in the beginning of the book about this being a journey for you. When people speak negatively about you, it is just another stone in the trail. Do not get fixated on it, do not let it turn you around or slow you down, it's just a distraction. Walk around it or past it and keep moving forward.

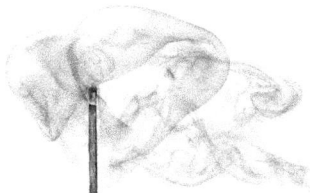

Chapter 27
LOVE WHAT YOU DO

Earlier, we discussed loving what you do from the perspective of customer service. Now I want to focus on you, your personal job, what you do in the firehouse, how you view your training and even the everyday events in the firehouse. I have pointed out previously how important love is because when you love someone or something you will do anything for them. This same principle applies to your leadership in the fire service. To be great at it requires loving it.

There is an important point for us to discuss at this time. When people have relationships, sometimes they think they are in love, but it is simply infatuation. What is the difference? When you are infatuated with a person, they are perfect in your eyes. You see no wrong or no negative qualities. That is not realistic. No one is perfect and no two people perfectly match in all things. Eventually you end up seeing the negatives, and the relationship ends—maybe painfully.

When you are in love, however, you see the person as they are. You can see the negatives, but the positives far out way them. There are areas where you don't see things exactly the same but the areas you do far outnumber them. When you are in love, you don't focus on the wrongs or negatives,

LEADERSHIP REFINED BY FIRE

you focus on the good aspects of that person. So, when a negative surfaces, you're able to honestly evaluate it, work to rectify it, or simply accept it. It does not destroy your love or relationship. When you love that person, you cannot imagine living without them. Their presence brings comfort to you and joy.

That is what I mean about loving what you do. If you have served more than several weeks in the fire service, you know there are negatives. There are people who sap the joy out of the job and make it difficult. In addition, the job itself can be very tough. Seeing children hurt, or die, seeing older ones pass away and watching their partner of 40, 50, or 60 years just helplessly mourning, seeing pain in fellow humans all the time can wear on you. The politics of the job, rumors, backstabbing, and other negatives are there in heavy doses.

When you experience these difficulties, these negatives, but can see past them because you love the job, you love the help you give to people, you see all the positives, then that is love instead of infatuation.

Sometimes, when kids are young, they see firefighters and long to be one. When they finally get the job, it's different than what they expected. They didn't know about the negatives, they only saw the glory, the camaraderie, the family atmosphere. When reality hits them in the face, it can be painful. If they haven't grown to truly love the job, to see its positives, sometimes they go down the wrong road and sometimes they leave.

I saw this happen to someone I was close to, someone I cared about very much. I watched as he came in bright-eyed, enthused, excited, and thrilled. He excelled for some time. But when the reality of the fire service hit him, when he began to experience the negatives, he began to experience cracks. It was obvious in his attitude right away. I tried to speak to him many times, but he took refuge with those who constantly spoke negatively about the department and the service.

Eventually, it led him to doing something stupid for which he was fired. For him it didn't matter, by that time he absolutely hated the fire department. What a sad event, but his story is not the only one that will end that way. When reality smacks you down, it can be difficult to get back up. This is true even as you promote in the service. Taking on new responsibilities, like a leadership role, brings with it all new challenges and negatives. However, the positives significantly outweigh those negatives. The privileges that become open to you can significantly alter your life in a positive way. It's all how you see it, whether you love what you do, you're just infatuated with it, or, worse, it's just a job to you.

Steve Jobs is credited with some incredible advancements in technology. Although he helped revolutionize the personal computer business through the company he cofounded, Apple, he was actually terminated by the company he helped create. He was eventually brought back to Apple on the verge of its collapse, and his love for technology and how technology can serve people helped propel the company forward to become one of the most innovative in history, revolutionizing the phone business, the music business, and now getting into the health business.[1]

Do you think you are really that different than Steve Jobs? All he did was work at something he loved. When we love something, we are able to put an incredible amount of time, effort, pain, money, and sacrifice into it. That is a recipe for success in any industry. He once said: "For the past 33 years, I have looked into the mirror every morning and asked myself: 'If today where the last day of my life, would I want to do what I am about to do today?' Whenever the answer has been 'no' for too many days in a row, I know I need to change something."[2] Ask yourself that question for the next month. If your answer is "no" for too many days in a row, then make a change, either *start loving what you are doing or start doing something you love.*

That does not mean you need to leave your current job but what it does mean is you need to leave your current state of mind.

If you never really fell in love with the job, if you were just infatuated with it, or if you were in love but have fallen out of love with it, how can you fix it? How can you get to the place where you love what you do?

SELF-EXAMINE.

As with every area in this book, it starts with self-examination. Why do you no longer love the job? Is it because it's different than you thought it would be? Did you expect too much, perhaps perfection? Is it because of the actions of others?

FOCUS ON THE POSITIVES THE FIRE SERVICE MAKES IN PEOPLE'S LIVES.

Think about what incredible service the fire service does for those it serves. How many lives have been rescued, how much pain has been averted, how much property has been saved? It is immeasurable. You are partly responsible for that. You are a part of that. What a great service to humanity.

FOCUS ON THE POSITIVES THE SERVICE OFFERS YOU AND YOUR FAMILY.

A job gives you an income that allows you to help care for your family, to fulfill your dreams and the dreams of those you care for. But an income is only a small part of the positives. The things you will learn on this job, this journey, will help you throughout your life. They will help your family and all those you care about. When my family and friends have had medical emergencies and ended up in the hospital or being treated by a doctor, it's always been a relief to me and the others in the situation that I have a good medical understanding—not so that I can tell the doctors what to do, but so that I understand their course of action and the direction they are looking. I can be a comfort and have been. This kind of training is especially helpful if you have or are planning on having kids.

PUT A GREATER EFFORT INTO THE JOB.

Have you ever heard the expression that you are doing something "just enough to hate it"? Coming at a tough job or assignment half-heartedly can make you hate it. However, if you commit, if you give it the full effort it deserves, it not only becomes easier but you might just enjoy it. Hose drills are one of the most hated exercises. Pulling, charging, dragging, draining, and repacking over and over again in the heat is tough and tiring. On one particular day, early in my career, my good friend was doing everything he could to avoid the work. It was blatantly obvious to the rest of us and I have to admit, I was tempted to do the same. Finally, one of the officers went over to him and said, "If you work as hard at helping us as you do trying to get out of the work, it would be a lot easier for you and the rest of us." We all just cracked up laughing. But it was a great lesson for me. We all started working a little harder. It not only made the drills go a lot faster, but it felt good, all of us working hard together to accomplish a task. I would recall that lesson several times in my career.

REACH OUT FOR NEW SERVICE OPPORTUNITIES, NEW AREAS TO TRAIN OR STUDY.

Nothing is more exciting than learning something new or attempting a new challenge. The fire service offers so many areas for you to specialize in, and many areas of study and operations to dig into. Try to become an expert in an area. Progress brings a great feeling of success, and when we take on a new challenge, it represents progress. Psychologist Mihalyi Csikszentmihalyi is the world's leading researcher of positive psychology—the study of what makes people truly happy. In Jonathan Haidt's book, *The Happiness Hypothesis*, he summarizes his research as follows: "Csikszentmihalyi's big discovery is there is a state many people value even more than chocolate after sex. It is the state of total immersion in a task that is challenging yet closely matched to one's abilities. It is what people sometimes call 'being in the zone.' Csikszentmihalyi called it 'flow' because it often feels like effortless movement."[2]

Get in the flow. Immerse yourself in a challenging new task where you learn something new.

REACH OUT TO OTHER FIRE DEPARTMENTS, MAKE FRIENDS, SEE WHAT THINGS THEY ARE DOING.

Being in a fire department is like family. You know that already. But the entire fire service is that way as well. Make friends in other departments. Learn from some of the things they are doing and bring them back to your department. Teach them the things you are doing in your department. Train with them, hang out with them. Enlarge your circle of friends.

READ FIRE JOURNALS AND BOOKS.

The fire service has many books and journals and magazines full of material to read and learn from. YouTube has a wide selection of videos of firefighters fighting fires, handling emergencies, and training. There are new tools to learn about and new techniques.

HELP OTHERS.

One of the greatest ways to experience joy is to help others. Can you share your knowledge with others? Can you help train others in the department? Have you ever thought of making training videos yourself for those in your department or even posting them on the internet? Doing so can open up a large family of firefighters to you.

GO TO SEMINARS AND CONVENTIONS.

When I had the privilege, attending conventions was among my favorite things to do. At these events you get to meet firefighters—and sometimes their families—from all over the country, and amazing friendships can develop. Attending seminars also gives you the opportunity to learn new techniques and skills while gaining deep respect for the fire service itself. You'll hear wonderful stories of heroism, brotherhood, love, and exceptional care, you'll see the greatness of the traditions we hold and the respect the country has for us, and you'll feel a part of this great organization!

Do you genuinely love what you do? Have you lost the love you once had? Are you at the point where you are thinking about leaving the job? Steve Jobs said that getting terminated was the best thing that ever happened to him. He was devastated at first and felt like he'd let the previous generation of entrepreneurs down. He said he was a very public failure and considered leaving Silicon Valley for good. But it allowed him to focus again, and something slowly began to dawn on him. "I still loved what I did. I had been rejected, but I was still in love," he said. He decided to start over. The freeness he felt allowed him to enter what he termed the most creative period of his life.[2] Focus. Follow the above steps and focus. You will find your love.

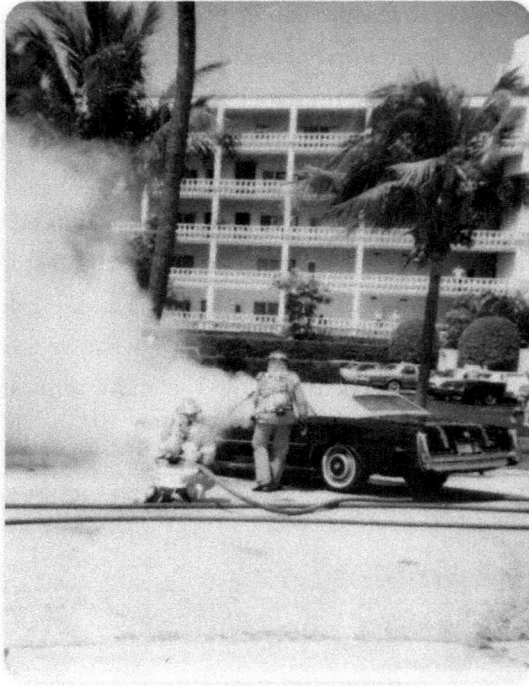

First year on the job. A car fire.
I am the one on the ground.

With my friend Jesse. We were checking out the ocean
as we waited for a Hurricane to make landfall.

During a bad storm, using a chainsaw to cut
a large tree that fell onto the road.

Cutting a vent hole in the roof of a house fire.

Looking up at the second floor of a house fire with my friend Mario. I was asking command to allow me to enter the house through the second floor window.

The doorway was blocked so I took my crew and entered this house fire through a front window.

A very young, skinny me with my Oakley sunglasses on keeping people away from what would end up being a dummy missile. Daily News Photo by Sig Bokalders.

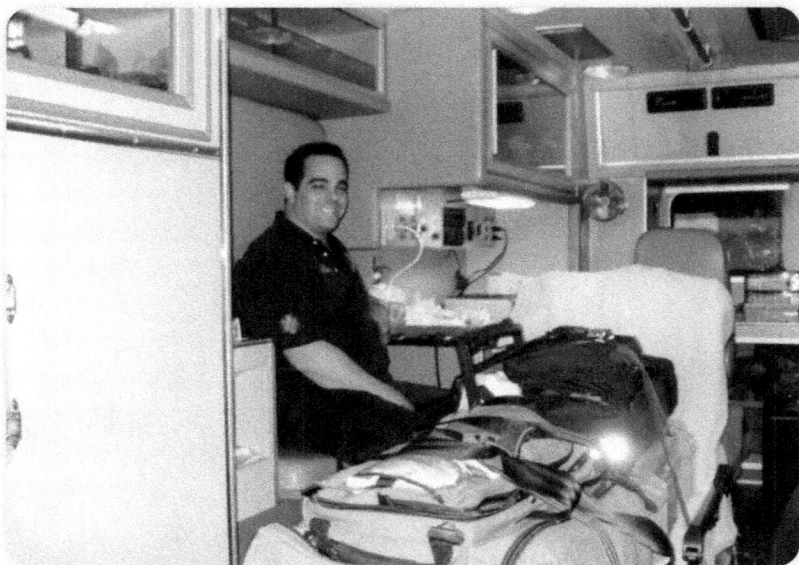

Checking out the back of the Rescue Vehicle.

Sunday morning at the table with my friend Eric. I
didn't eat breakfast but stayed with the crew.

At the table with my friend Mario. My book on
Major Dick Winters is on the table.

My wife and I holding our kids at one
of my promotional ceremonies.

Early in my career in my dress uniform.

With my boys when I was promoted to Lieutenant.

My boys playing with the blood pressure cuff
in the public room at station 3.

With my wife and boys in the training room.

With another one of my loves, my 1992 Mustang
convertible GT, Paxton supercharged. I miss that
car. One of the guys wrote Elvis on it.

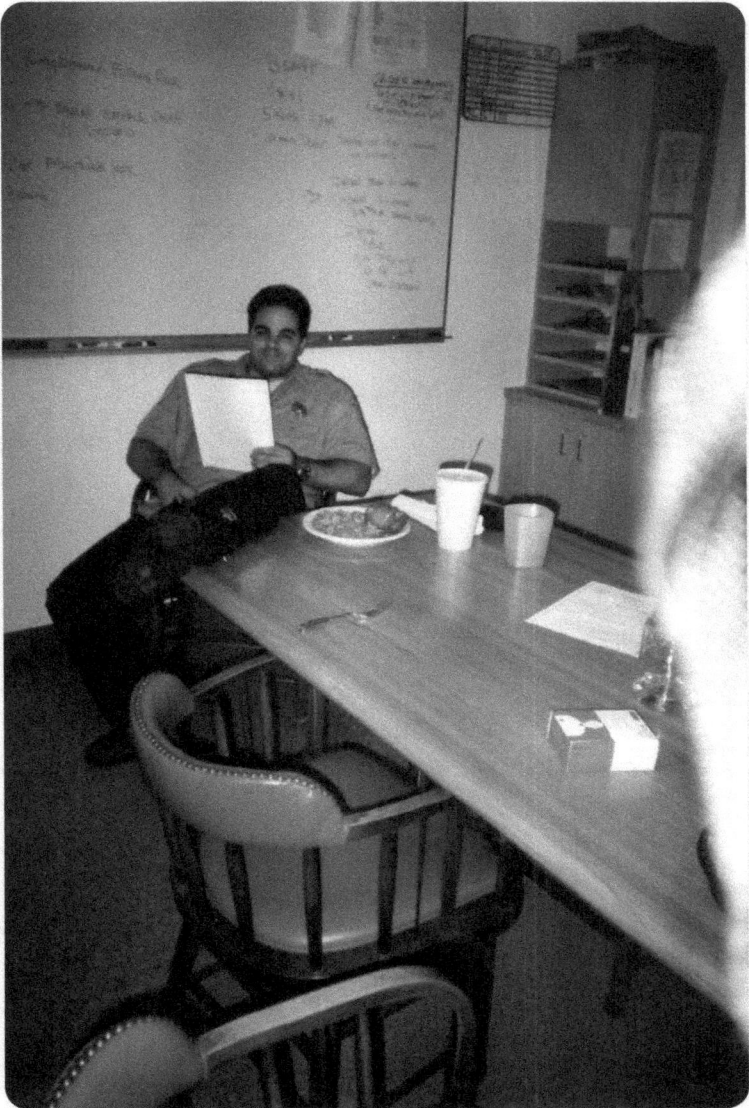

Sitting at the table, studying while I eat
what looks like a stuffed pepper.

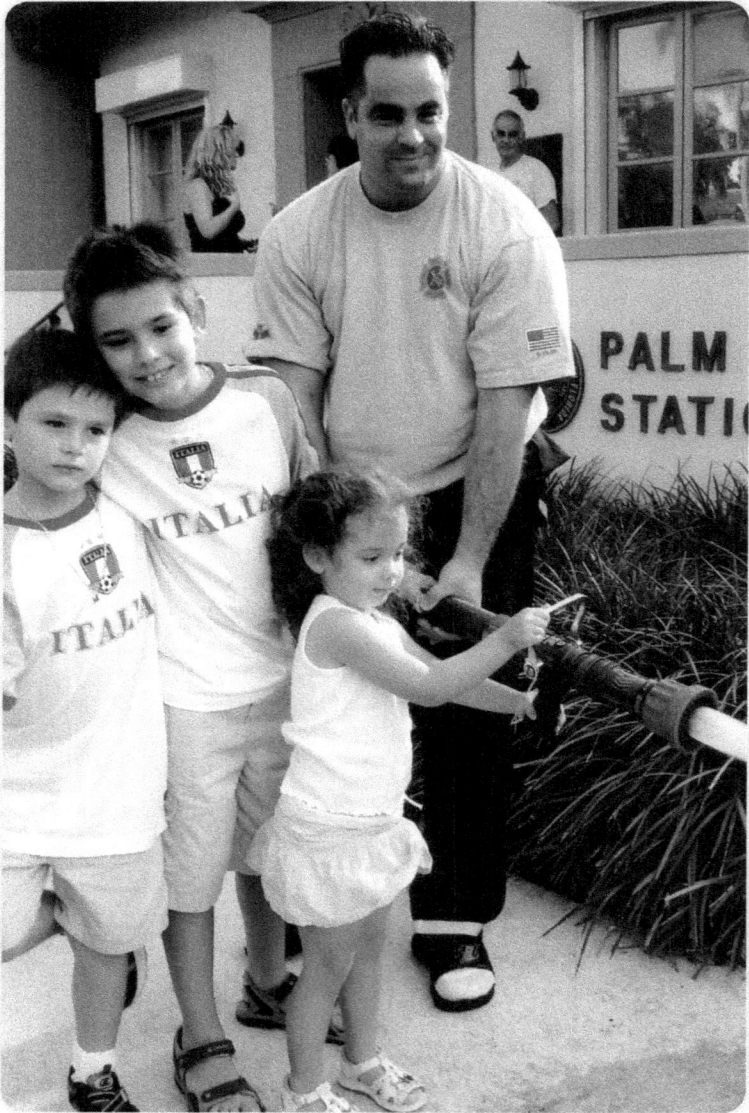

My two sons and my niece Alexa flowing the fire hose.

My niece Gianna in the fire engine.

When I first received my bunker gear for fire school. That
gear was terribly old. Holding my girlfriend who would about
a year later become my wife. I didn't even know how to put
my gear on yet, notice my boots were outside my pants.

Picking up the love of my life. After we got married
and before the reception we went and took pictures by
the fire truck. I absolutely loved the fire service.

My son Matt in the front of the Engine. We put
together a whole program for him and his class.

Teaching a class and showing them the Engine
with my friends Joe, Mario, and Stuart.

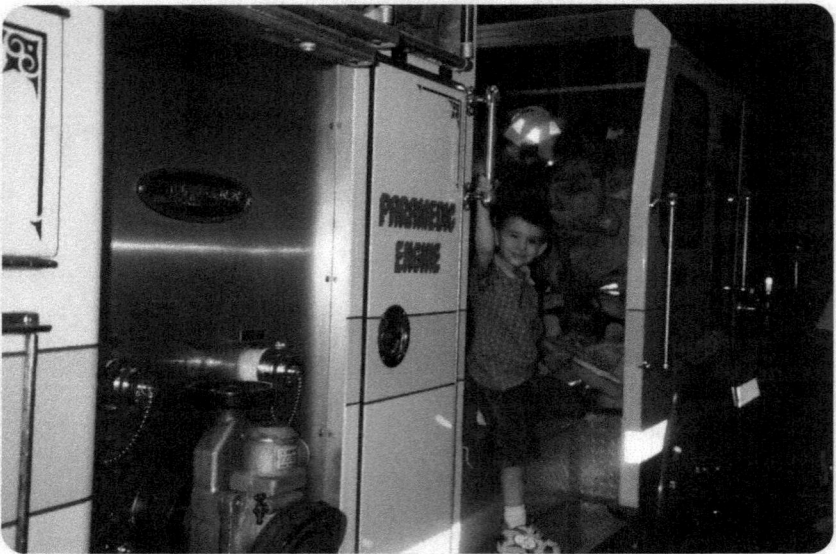

My son Mike checking out my backseat in the Engine.

My Dad and my brother at one of my promotion ceremonies.

My friend Stuart holding up a toy we collected for the "Toys for Tots" campaign the department participated in each year.

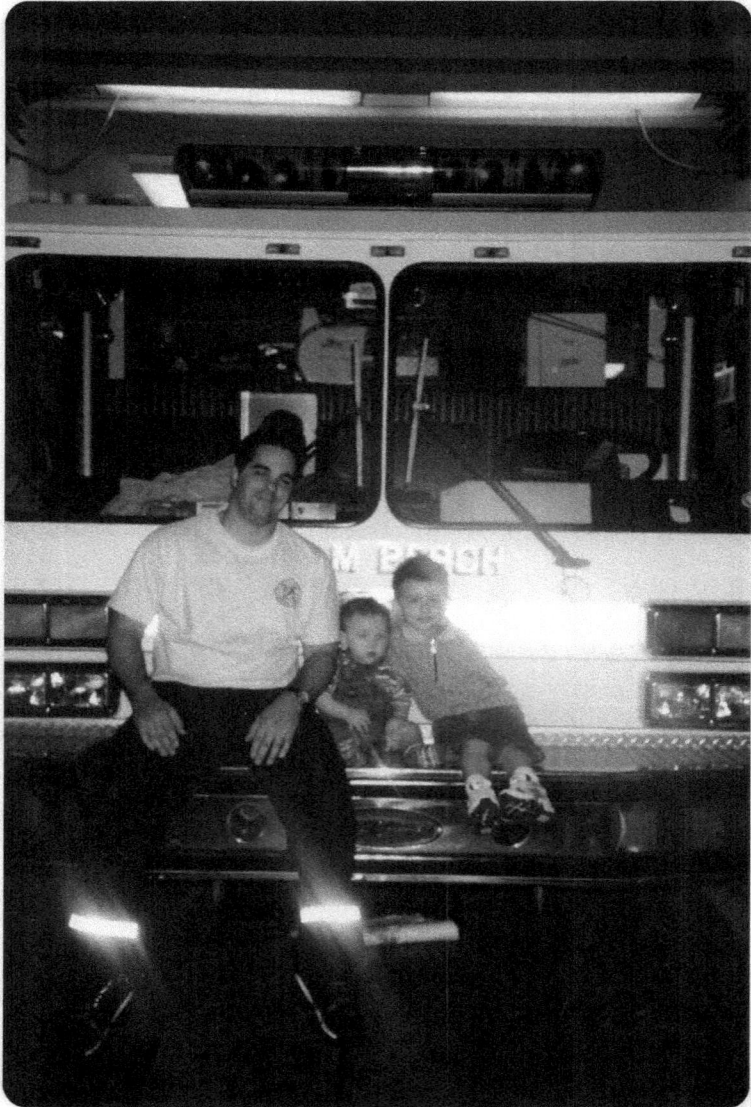

Sitting with my boys on the front of the Engine.

Training- working with different tools to remove different doors.

One of the many houses and buildings we smoked out and trained in.

My friend Mario, wearing a blacked out
mask, goes through a hose drill.

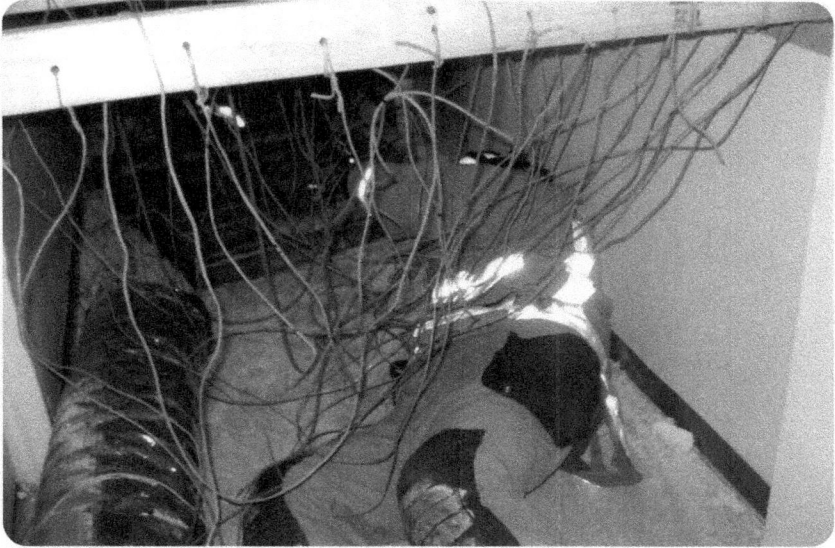

Training in an entanglement prop we built in
a warehouse in west palm beach.

My friend Stuart getting ready to go through the entanglement prop.

Vehicle extrication training. Level II. Large Vehicles, trucks and buses.

Stabilizing a bus in training before we tore it up.

Demonstrating how to get through a small opening
while bunkered out, on a prop I built.

Demonstrating how to get through a thin opening
while bunkered out, on a prop I built.

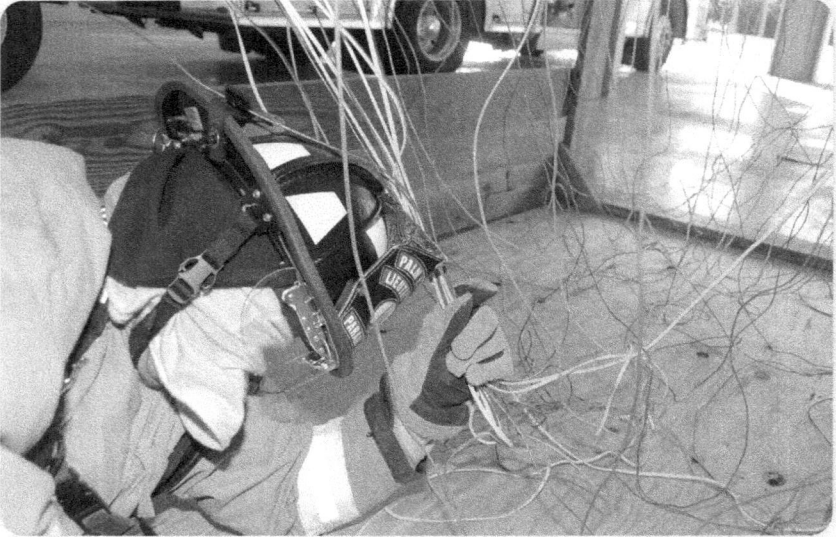

Demonstrating on an entanglement prop I built at the station.

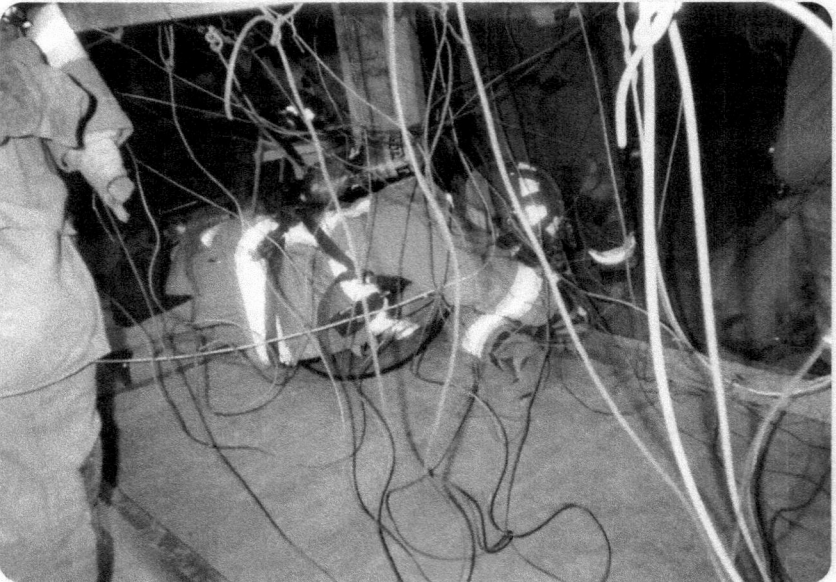

Going through the entanglement.

Practicing victim carries on my friend Joe up the stairs.

Preparing a victim to be lowered from a multi-story home.

Working with chainsaws in a training drill.

Coming out of a window during a training class.

With my wife Kim on the last day of my career.

With my boys on my last day.

My room the last night of my career.

Emptying my locker and leaving after my last shift.

Sitting with my niece Kayla on Engine 2.

My son Matthew and my nieces Kayla and
Kristen in the kitchen at Station 2.

My nephew Nick and my boys playing cards.

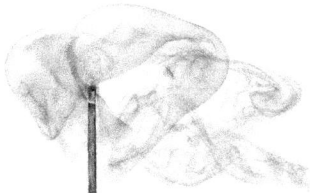

Chapter 28
COMMUNICATION

*"Language!—the blood of the soul, Sir! into which
our thoughts run and out of which they grow!"*
— Oliver Wendell Holmes Sr.[1]

The power that language or communication possess cannot be overstated. It is language, or the ability for people to communicate with each other, that has had the greatest impact on mankind's progress. The Bible beautifully illustrates this. In the Book of Genesis, chapter 11, the story of the Tower of Babel is told. It speaks of when people on Earth all spoke one language. Their desire was to build a celebrated city for themselves. In the story, God came down to see what they were doing, and expressed the following, "If as one people speaking the same language they have begun to do this, then nothing they plan to do will be impossible for them. Come, let us go down and confuse their language so they will not understand each other."[2]

The story captures the power of communication.

Notice that the key to halting their progress was confusing their language, otherwise nothing would have been impossible for them to accomplish. On a macro scale, it's obvious that communication skills are necessary, but what about on a micro scale—in the department—for a leader?

Obviously, we have to be able to understand each other, to speak the same language. But is that all there is to it? Nitin Nohria, dean of Harvard Business school, writes: "Communications is the real work of leadership."[3] Remember that leading is about getting a goal or objective accomplished. The reason for leadership is because others need to be brought in and guided to help accomplish the objective. So, when you think of communications, have in mind the necessity of including others. Let's consider a couple of important aspects.

1 - Be as open as possible.

Always try to be open with everyone around you as long as you are communicating information that can be shared. Communication goes both up and down the ladder. For those you lead, do the same. Be forthright, giving information and receiving it. People intentionally withhold information usually for two reasons. First, because they want to be the one in charge and having all the information makes people rely on them. Second, they are unsure—or flatly don't really know what to do—and they don't want those they lead to know that. Sharing the game plan, your thoughts, your ideas, and problems with those you lead will help them feel they are part of the team. Do not keep people in the dark or guessing. If you do, they will feel as though you don't trust them. Hence, they will withhold trust from you as well.

Discretion is advised here. There is no need to explain every bit of minutia involved as this will make you appear ineffective or weak. You still are the leader and that must be understood and there will be things that you may not be able to share.

There will be many times when you may be unsure of how to handle things. Asking for input—not direction—from those you lead can have several very positive effects. This is an excellent way to train those you lead to begin to use critical thinking skills, especially at this stage when their decisions won't cause a problem. Helping them to learn to process and formulate a strategy and learn from any mistake they make in this process will give them good footing for their own future role as a leader. Discussing strategy and events together and even following up with breakdown and analysis can be great training for someone in their position, helping them to see the bigger picture, so that when they are promoted, they have some experience with working on creating a strategy and leading that strategy.

Doing this also gives you an idea of their strategic thinking ability and where they may need additional training. This will help you to focus your guidance on that area. Additionally, it will draw them closer to you. Whenever a person feels like someone is taking an interest in them and helping them progress—especially their boss—it builds gratitude in them and makes them feel special, an important part of the team. Of course, you never want to do this as a testing situation, where the person feels pressure to perform well in something that they have not been trained in.

There are also personal benefits to a strategy of open communication. We don't always know everything, nor do we always see things the right way. Having others express their thoughts on a situation can give you a different perspective, which can help you decide on a direction.

2 - Help those you lead to progress, build them up, and never tear down.

Let's highlight two areas. First, look for opportunities to get those you lead involved with the direction the department is going. Discuss with them the goals of the department and the goals set out by you for the crew. Get them involved with that direction, thinking that way, looking for ways to improve in that direction and help the department. There is nothing more encouraging for a firefighter, or anyone really, than to feel a part of the team, someone who counts, someone whose input is valued. Use the time in between evaluations to have discussions about their goals and their progress, how they are doing, how you can help them, and what they can do to continue moving in a forward direction.

Second, never tear them down with words. In the fire service, talking negatively about people is a daily event in many crews. Putting people down, mocking their faults or their deficiencies is commonplace. And it is wrong. I try not to say too many things are definitively wrong or right, for who am I to tell others the right way to live? However, I have always felt strongly about this. I don't believe in hazing; it seems to be something that bullies love. I have been told that my view makes people weak. I disagree, the only weak people I see are the ones making fun of others. Making fun of others doesn't make people strong, it causes divisions, it tears people down, it isolates people, and it causes distrust.

Instead, look for opportunities to praise, to point out excellence, to tell others about the good things your subordinates are doing. When it comes back to your subordinates that you have been bragging about them to others, what do you think they will feel? How much trust will be built in your leadership? How hard will they want to work to do even better in appreciation for what you said and because they feel better about themselves and what they can accomplish? It is a win-win for you.

3 - Speech about the department, others, and rumors

Remember that leadership is accomplishing whatever goal or task you are presented with, using the team you have, making those you lead better, and making them excel for the sake of the department and those we serve. You cannot do that when you disparage the department or others or engage in rumors. We will discuss this in more detail in the chapter on rumors.

4 - Difficult conversations

There will be times when conversations may be extremely difficult. Sometimes these conversations can get heated. How do you handle that?

If time allows, try to prepare. Sometimes we can't because things just pop up, however if you are able to postpone a conversation, tell the person,

"We will discuss these matters in an hour or so." This will give you time to think and prepare.

Think about what you are going to say. Envision the end result: where you want to get to. Have a positive goal in mind. This will help you to stay focused. Difficult and heated conversations can veer off course. They can open a whole set of other issues. Try to stay focused on the direct issue.

Stay calm. Resolve in your mind to be calm. If you get heated, it never helps the situation; instead, it does one of three things. Getting heated can intimidate the other person, causing them to shut down during the conversation. It can also escalate the problem, perhaps even adding fuel to an already heated subject. This can cause things to get worse than they were before you started. Finally, if you get heated and they stay calm, they may feel they are in better command of their emotions and will stop trying to converse. They will see you as unreasonable and lacking in leadership.

A cool head and soothing tone can quickly calm the situation. Even if you are unable to completely resolve the situation or problem at that moment, in that conversation, remaining calm will leave the door open for a future conversation, as the person will trust you even more with their feelings.

Remember why you are there. Remember that as a leader, you are there to help the person be better. When you go into the discussion with that in mind, it will keep you from taking things personally or getting heated yourself. I went on a call where an 82-year-old woman called me every curse word in the book. She spat at me and tried to hit me. As bad as it sounds, it was actually quite comical.

Anyway, we administered her some glucose and in 60 seconds the sweetest 82-year-old lady appeared. Her first question was, "What is going on?" She was so gentle and kind. From your experience in the fire service, you probably guessed that she was a diabetic and her sugar had dropped way too low. I have had many of those experiences as I am sure you have as well. The point is, I knew why I was there. When we tested her sugar and it was low, and she had a history of being a diabetic, the focus was on getting her better. It didn't matter what she was calling me. In the end, I helped her. Remember why you are having the discussion in the first place.

Sincerely listen. Let them talk. Just listen and then try to repeat back what it is that's upsetting them. You do this not to placate them, but so that *you* understand how they feel and they feel understood by you. You may be surprised how well that works and how fast it can diffuse the situation. There are three reasons for this:

1. People want to feel that they are understood. When you take the time to genuinely listen, and they can see that, it will help them to open up.

2. Repeating back to them makes them feel understood. When you say, "So if I understand you correctly, you feel..." and your analysis is correct, they will acknowledge that. Christopher Voss, a former hostage negotiator for the FBI, who wrote the book, *Never Split the Difference: Negotiating As If Your Life Depends Upon It*, advises, "Once you have articulated their perspective for them, they feel understood. And a person who feels understood is getting a feel-good wave of chemicals in their brain. The one you are really going for is oxytocin, the bonding chemical. Once they get a hit of oxytocin, everything is going to change. They feel bonded to you."[4] And if they feel bonded, whether it's a little or a lot, it will help resolve the issue in a calmer manner.

3. Finally, when you genuinely and sincerely listen to someone else's perspective, you may change your own mind on how you feel about a situation. You may see that you were wrong or could have handled the situation better, or perhaps the situation wasn't as bad as you initially thought.

Words are singularly the most powerful force available to humanity. We can choose to use this force constructively with words of encouragement, or destructively using words of despair. Words have the power to help and to hurt, to build up and to tear down, to fix and to destroy. One way or another your words will have an impact. You must choose which way. Be free and generous with information to those you lead and those who lead you. Information makes people part of the team, part of the plan, and part of you. Invite that trust and give that trust and you will be rewarded.

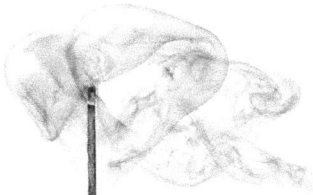

Chapter 29
POLITICS ON THE JOB

When I speak to firefighters and police officers about the job—and it doesn't matter which department or even which state they're in—one of the biggest negatives they mention is the "politics" on the job. It cannot be denied that over the course of the last couple of decades our society has become very politically oriented. Politics has worked its way into every facet of our social experience. And although all over this great country the conversations in the departments can turn to heated discussions about Republican and Democrat policies, that is not what firefighters and police officers are normally referring to when they talk about "politics on the job."

Politics on the job refers to the unequal treatment of people based on who they know. Individuals get treated differently, have different opportunities presented to them, get better raises, and nicer assignments based on their relationships with other individuals in the department of higher rank. Does this sound familiar?

My goal in writing this book was to bring something positive to individuals, not to focus on the negatives. However, this must be addressed as it not only permeates most, if not every department I've ever had knowledge of, it can be debilitating to the career of those on the wrong side. It's imperative that you personally never engage in this behavior.

So, let's discuss it. The fire service is an organization filled with rules, regulations, policies, procedures, and a host of other directives. It's extremely easy to break one of these sooner or later. Trying to do your job while walking on eggshells, wondering which of these hundreds and hundreds of directives might be used against you, can completely stymie your progress with moving forward, bettering yourself and bettering the department. So how do we work through the politics on the job?

SELF-EXAMINE

As usual, start with yourself. When presented with an event where you feel you were the victim of an injustice simply because of the "good old boy system" or because you didn't know the right people or because you feel you are not liked, take a moment to truly examine the entire situation and your actions in it. Did your actions, attitude, or work ethic have something to do with it? Are you seeing the situation clearly? Or are you convincing yourself of something that isn't there?

In his book, *The Happiness Hypothesis*, Jonathan Haidt, examined the work of psychologists Emily Pronin from Princeton and Lee Ross from Stanford on a phenomenon called naïve realism. He states, "Pronin and Ross trace this resistance to a phenomenon they call "naïve realism." Each of us thinks we see the world directly as it really is. We further believe that the facts as we see them are there for all to see, therefore others should agree with us. If they don't agree, it follows either that they have not yet been exposed to the relevant facts or else that they are blinded by their interests and ideologies."[1] Is it possible that *you* are not seeing the whole picture?

Friends of mine asked me to talk to their son. He had been hired as a firefighter for a neighboring city about two years earlier and was now very unhappy. When I spoke to him, he said that he had been extremely happy and had a great crew and they all got along—then the chief moved him to another shift. He missed his friends and the new crew didn't get along nearly as well. He was extremely jaded. I tried to get him to see it in a different light. I explained to him that he was still very new and should take any assignment without complaining, how important it was for the administration to see him as a team player and not a selfish one. He didn't know the previous crew before he got hired, I expounded, but became really good friends and he could do the same with the new crew. In addition, I gave him an example of something similar that had happened to me. Nothing I said worked. He was angry and jaded.

Six months later I bumped into his mom at a grocery store. I asked about her son. She told me he wasn't doing well at all. Several months after the shift change had taken place the chief called him into his office. The chief told him that he had been so impressed with him after he hired him and so appreciated his team mentality that he moved him to the new shift so he could help that crew. The chief stated he was hoping that he would be able to help that crew come together, but that instead he brought a terrible attitude with him and made it worse.

His mom told me that he was now extremely depressed, that he realized how wrong he was and how he blew an excellent opportunity to show the administration that he was not only a team player but that he could be counted on by the administration whenever they needed him. That kind of reputation is hard to build and even harder to get back once lost.

It is easy to get lost in our own perception of things. Do a thorough self-examination and perhaps even rely upon a trusted friend or mentor.

DON'T LET IT DERAIL YOU

Almost nothing can feel worse than being unfairly or unjustly treated. It is debilitating. It can crush your dreams, your career, and even hinder you in your personal life. It can make you recede from pursuing any new opportunities and even become a recluse in the department. I have seen it happen to others. I have felt it personally. That has to be the exact opposite reaction you want, especially as a leader.

Early in my career, I identified a lieutenant who I thought I could learn from. He was brilliant, a very quick thinker, fair, and kind. The men and women all respected him. In addition, they had voted him to be the union president. He reached out for promotion to captain and scored number one. In fact, the distance between him and the number two guy was huge. He crushed it. However, he was passed over and the spot was given to the number two guy. That year the city had decided not to give the firefighters a cost of living raise but, instead, a one-time bonus. The department felt hurt by this and, as union president, the lieutenant had written a letter to the city officials stating such. The chief used that act against him and passed him over. He was crushed. He couldn't even talk about it. His eyes would water up whenever it was even mentioned. His once gregarious and bubbly personality became cool and withdrawn. But he came to work, did his job and did it well, and never spoke a word about the incident.

After about six months, another captain, who was disgusted by what happened to that lieutenant, decided to take an early retirement so that the lieutenant could be promoted before the list expired. Thankfully, the chief promoted him this time. Would he have been promoted if he had gone around and complained about it? Or if he decided he was going to do a terrible job from then on? Certainly not. He received what he should have gotten initially, and he didn't complain or let it affect his job performance. Instead, he was viewed as stoic and strong. The respect from the men and women only grew.

Even if he hadn't been promoted off that list and he had to compete again, as much as that would have been tough, that is what you have to do. You get up and get back on your horse. You don't stay down when you trip, and you don't let other people's actions change or thwart your desires and goals. You work hard to put it past you. You examine yourself and see if there is something you can do that helps you even more the next time— and then you get back in the ring and compete again. As a leader, everyone is watching how you will respond to this adversity. You will set an example one way or the other.

DO NOT BECOME WHAT YOU HATE

Becoming what you hate is easy when you start to see the hierarchy. To see the politics. To see how some get treated differently than others. And it's easy to follow the same path. To sell your soul to those who can help you. I beg you not to go down that road. A sycophant will never be a leader and a leader will never be a sycophant. We must make a stop to this type of behavior.

We must get people into places of authority who succeed because they did things the right way. Otherwise, individuals who are not hard working, who try to cheat the system, who are not good firefighters, achieve these spots and the terrible cycle continues. To be sure, not all individuals in position of authority got there unqualified, I am not saying they have. Most have worked very hard, however some have not. Our job requires physical and mental strength as well as a heaping dose of courage. When you take the easy road, you tell yourself and others that you lack strength, courage, integrity, and self-respect. These things will tear at your soul and destroy you from the inside. They also make you appear weak and cowardly, and who wants to be led by someone they view that way?

Not participating in department politics means you don't allow the people you lead to influence you with sycophantic behavior either. Here too, it is the harder road. Remember that if someone is resorting to that behavior, it may be because they are not fully qualified to do the job, or they may be doing it because they believe it's easier. Either way, you don't want to promote that type of behavior in your department, it's detrimental.

In addition, you may have subordinates who could be fantastic for the department. If you reward bad behavior, you could lose the people who are willing to work hard to learn the whole job, become proficient at it, and reach out on their merits. We don't want to lose these men and women.

Unfortunately, we will all be faced with the detrimental effects of politics on the job. When it comes around for you, do a good self-examination and work on any issues. Do not let the politics derail you, get back in the ring, and don't become what you hate.

Imagine there are no politics in the department. It's easy if you try!

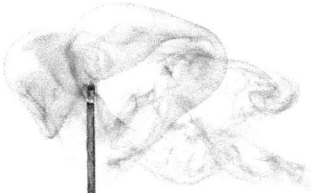

Chapter 30
POSITIVE ATTITUDE

Have a positive attitude. It sounds easy, but it's not. It's extremely difficult, especially because our careers last 20, 25, 30 years and even longer. Having a positive attitude for the long term in this crazy, busy, mixed up, stressful, political, painful world is far from simple when there are all those stresses of the job weighing on you as well. I think the importance of a positive attitude and how to work toward it deserves a deeper look.

The lack of a positive attitude can have a huge effect on your life. I worked with a firefighter who could not have been more negative. Do you remember Eeyore from *Winnie the Pooh*? No matter what good thing was happening, Eeyore found something to complain about. Well, this firefighter made Eeyore look like Mr. Positive Sunshine. It was terrible working with him. When he was assigned to a crew, the team was miserable.

I actually liked the guy and felt badly for him, but no matter how hard I tried to engage him in positive conversation, it was not possible. He could make paradise seem like a hellish nightmare if you were stuck with him there. Ultimately, he quit the department as angry as a human can be, and that's one effect of not having a positive attitude, but science has shown us the effects are even deeper.

Is a positive attitude really that important? If so, when? How do you remain positive? And how does being positive conflate with leadership? Let's discuss these questions.

What is a positive attitude? It can be rephrased simply as positive thinking, which "does not mean that you stick your head in the sand or view the world through rose colored glasses. Positive thinking encompasses the mental attitude of optimism, which *searches for favorable outcomes in all situations.* It relies on the emotional state of hope, which looks past the current circumstance and supports the building of emotional, social, and other resources for positive action."[1]

As much as we should try to remain positive all the time, obviously, that is not possible. However, we can begin each day with a positive attitude. We can wake up each day and make our first thoughts positive. Some days that's very easy—like the day you get married, or the day your child is born, the day you get promoted or get a raise, or it's Sunday and all the kids are coming over for pasta and meatballs. Those are great days. However, a positive attitude will have its best influence on the days when things are not going so well.

Research indicates that positive emotions contribute to important downstream life outcomes, including friendship development, marital satisfaction, higher incomes, and better physical health. People who experience frequent positive emotions have even been shown to live longer. The daily experience of positivity compounds over time to build a variety of consequential personal resources. This is called the "Build Hypothesis," which "holds that positive emotions set people on trajectories of growth that, over time, build consequential personal resources."[2] In addition, they "experience or express positive emotions more than others, and over time show increases in optimism and tranquility, ego-resilience, good mental health, and quality in their close relationships."[3]

Positive emotions broaden your sense of possibility and open your mind to more options, and the benefits don't stop after a few minutes of feeling good. In fact, *the biggest benefit is an enhanced ability to build skills and develop resources for use later in life.* It is like having additional tools in your toolbox for difficult projects, or extra money in a bank account for when financial problems hit. When you are experiencing positive emotions, you will see more possibilities in your life. On the other hand, negative emotions narrow your mind and focus your thoughts.

When you are stressed about everything you have to get done in a day, you may find it hard to actually start anything because you're paralyzed by your long to-do list. Negative emotions prevent your brain from seeing the other options and choices that surround you.[4]

And what about physical effects? Studies from Johns Hopkins University have shown that, for people with a family history of heart disease, those with a positive outlook were one-third less likely to have a heart attack or other cardiovascular event within five to 25 years than those with a more negative outlook.[5] Other studies have found that a positive attitude improves outcomes and life satisfaction across a spectrum of conditions—including traumatic brain injury, stroke, and brain tumors.[6]

In addition, children—specifically, your children—can be affected by the example you show them. Researchers at Stanford University School of Medicine studied attitudes in children ages seven to ten and used MRI brain scans to map the neurological effects of positivity.[7] Their findings showed that positivity enhanced children's problem-solving abilities and increased their memories by improving functions of the hippocampus— the area of the brain responsible for memory—along with other benefits. Being positive made them more successful.[8]

These examples are just a small taste of the information out there showing the tremendous benefits of having a positive attitude—you could fill a book with examples.

So, how do you maintain a positive attitude? Well, this book is all about taking personal responsibility for your life, your actions, and the example you set. It's about setting goals to better yourself and accomplishing those goals. So, start there, set a goal to have a positive attitude each day.

When you do it every day, it gets easier, and then it becomes natural. Here are a few steps to help.

1. **Make it a habit**. Make your decision to have a positive attitude a habit. Make it something you not only believe in but that you believe to the core of who you are. Work on it every day. Do not give up. Be especially diligent during difficult times. Ask those close to you if they perceive you as a positive person and if they don't, ask them to point out where you can improve. Do a good and honest self-examination. And do it often.

2. **Eliminate negativity**. Tony Robbins told the story in one of his talks that he once picked up a booklet called *The 7-Day Mental Diet* by Emmet Fox. It challenges readers to truly live for seven days without one negative thought. Robbins said that, although he had many difficulties going on in his life at that time, he decided to give it a try. His conclusion? *"It changed my life."*[9] He explained that after seven days you feel like a different person and people treat you differently.

    I used my Ben Franklin book chart, that I referenced in chapter 4, to help me give it a try. The challenge is you have to *complete* seven straight days without negativity: if you fail, you have to start over. At 2:00 p.m. on the first day I failed. I started over the next day. Two days in, I failed again. Like Tony Robbins at the time that he discovered this book, my life was going through some very difficult challenges. I thought momentarily that I should put the challenge off until things got a little more concrete. But then I realized that this was the very best time to do this challenge.

    I started again, and this time, I finally succeeded. It truly was life changing. What I discovered was that *in order to succeed at the challenge, I had to be in constant conscious control of what I was saying and thinking*. At first this is very difficult, but it gets easier, and when you master it, it feels like a beautiful cleansing and is fantastic for creating a positive attitude. We'll return to this subject in Chapter 35: "When the Going Gets Tough."

3. **Don't be judgmental**. Finding fault in other people will bring you down. If you're a fault finder, it won't end with others: you will consciously and subconsciously find fault with yourself as well, which will make you miserable. The way you judge is the way you will be judged. If you judge others and never give them the benefit of the doubt or try to understand what they might be going through or why they did what they did, others will do the same to you.

Try to empathize. My wife reminded me of a well-known quotation. "Everyone you meet is fighting a battle you know nothing about. Be Kind. Always." It's a beautiful thought. Don't judge, be kind, it will make you more positive and it will come back to you.

4. **Meditation**. Meditation has a whole host of benefits, both physical and mental—relaxation, balance, gratitude, refreshing the spirit, and creating a positive attitude. We will further discuss the benefits in dealing with PTSD later in the book. We're blessed to have many experts who teach and write about meditation and who can teach you how to do it through their books and internet videos.

5. **Write/Focus on the good**. Writing and journaling have gained significant popularity in recent years. Writing specifically about the positive things that are going on in your life each day, week by week, allows you to keep your mindset. Focusing on the good helps you to become positive and maintain a positive attitude.

One important thing about journaling this way and keeping a written record is that it allows you to go back during very difficult times to remember all the positive things that have happened in your life.

If you write for a while and stop, studies have found that even months later, people who journal this way have better mood levels, fewer visits to the health center, and experience fewer illnesses than those that who don't.[10] And if you don't want to write, just focus: Each night think of at least three positive things that happened to you that day and how they benefited you. This too will have an effect on our outlook.

6. **Association.** We discussed birds of a feather earlier: those you associate with will most definitely have an influence upon you. Try to find people with positive attitudes in your department to associate with. Set a fine example and you may be surprised. You will attract others who were looking for the same thing and may not have found it yet and in addition you will help others too, to seek good friends with positive people. Try meeting others on other shifts or taking classes with firefighters from other departments. I have met many brilliant, enthusiastic, and positive men and women in fire classes and got to know their circle of friends as well.

7. **Enjoy the gift of life**. We are such blessed people. Of course, we face many difficulties and sad events too, but life is fantastic and being a firefighter is the greatest job in the world. Our job puts a lot of stress on us and stepping into a leadership role has even more stress and pressure, which is why it's so important to enjoy the gift of life. Explore the beautiful world we have. Get to know cultures all over the globe. See the Grand Canyon, relax on a beach, scuba dive, watch a sunset or sunrise, smell the flowers. Make sure you are taking time to enjoy this precious time you have.

How does a positive attitude conflate with leadership? Remember, as a leader, others follow *you*, not because you are in charge, or because you can discipline them if they don't, but because they *want to*. They follow your directions and learn from your actions and positivity.

Having a negative attitude creates divisions in the department, destroys morale, kills training, destroys trust, creates a hostile work environment and so much more.

A positive attitude does the opposite. With a positive attitude, you can overcome difficult times and issues in the department, draw the crew and department together, and lead them to the next level. You will have a significant effect on those you lead and all those around you. What could be better?

Have a positive attitude!

Chapter 31
PRIDE

In your mind what is the difference between good art and great art? Stop for a moment and consider that point. Leo Tolstoy, in *What is Art?*, wrote, "A real work of art destroys, in the consciousness of the receiver, the separation between himself and the artist."[1] That is a powerful statement. Great art makes you feel like you are a part of it, like it isn't art but something that resonates within you.

What does that have to do with pride? Pride is one of those words that has two almost opposite meanings. We are not talking about the kind of pride that is defined as "a high or inordinate opinion of one's own dignity, importance, merit or superiority, whether as cherished in the mind or as displayed in bearing, conduct, etc."[2] That is a negative pride that we never want to display. When we speak positively of pride, we are talking about taking pride in all we do. That is pride defined as "a becoming or dignified sense of what is due to oneself or one's position or character."[3] That means making everything you do, everything you work on and produce, and everything you are, stand out as having a standard, something you can always be proud of.

What does this have to do with art? Well art and all other things produced by someone can be of varying degrees of excellence. Mr. Tolstoy showed that great art makes a connection with the person. The artist ultimately made a tremendous effort to make that connection with the viewer, fixing, refining, clarifying details until he could show his product with pride. That is what we are speaking about. In every effort you put forth, make a connection with those it affects so that it has an impact on their lives.

When you do a training with your firefighters, do they feel that your efforts connect with them? Do they see it not just as training but as something beneficial, something that resonates with them? When you interact with the customer on an emergency does your customer service transcend what they were expecting so that —as Tolstoy put it—it destroys, in the consciousness of the receiver, the separation between himself and you? Does the customer feel totally in your care? Do they feel there is a connection? When you lead others, does your pride in your work have you concerned even with the details so that those being led feel the connection of a true leader? This is the goal.

How do we display this good pride? When you work on a project, whether it is cleaning the fire engine, putting out a fire, helping a patient with chest pain, or just answering a question from the public, do it in the best possible way it can be done and in a manner that dignifies you, your service to others, and the fire service itself. Make this the quintessential you, how you do all the things you do.

Putting pride in all I do received the harshest test I could have ever imagined. The 2008 economic difficulties caused the city I served to cut almost every benefit we had. We lost our pension and went to a 401(K). We lost our Kelly days. We were expected to work an additional six hours a week for the same pay. We were unable to use sick time without punishment. We went about eight years without a raise. We were no longer allowed to shop for food during shift time. Swaps were made more difficult. And there were a host of other benefit losses. What made matters worse was that the non-union administration-level personnel did not suffer all of these changes at all. They even kept the Kelly days. On top of that, we were a small city, and all the surrounding cities in our area made either no changes or extremely minimal changes.

The morale in our department sank to lows I had never seen. People were full of anger and felt completely betrayed by the city and by the fire administration. People's family lives were affected with divorces and self-deterioration. Some people were terminated, and others began leaving either to be hired by other local fire departments or to get out of fire service completely. This left us low on personnel and forced overtime, which only magnified and accelerated the problems. I was a captain and stepped up as battalion chief often, whenever one of the battalion chiefs was off on their Kelly days, vacations, or sick.

Many firefighters began to flatly refuse to do station duties or train the right way. Many felt that city residents did not stand up for us against the city leaders and they began to give terrible service.

I too was affected. I was hurt and felt betrayed by the city officials and the administration, but I refused to let that affect my pride in what I did and my pride in the fire service itself. Some of the team mocked me and assumed I was just trying to please the administration for a future promotion. Nothing was further from the truth. Nothing. How could I ever want to be a part of an administration that I felt had betrayed me?

But I persisted, and it wasn't so easy pushing my crew to do the right thing. I explained to them, and to members who would work with us at times, that this negative turn of events would not stop us from doing the best job we can, from excelling in customer service and shining as a Fire Department. There were many days when I questioned myself. Was I doing the right thing? But I look back now with pride about the way I handled that terrible situation.

I loved being a firefighter, I loved serving the residents, I loved leading, and nothing anyone else could say or do would change who I am inside. It made a difference in the lives of those I led within the department during that time, it made a difference in those we served in the city, and it made a difference in me, for the rest of my life. I would not change that feeling for anything in the world. It has propelled me to continue on the right course no matter how difficult the circumstances. I am grateful for the challenge and the way I handled it, because the pride I felt for what I did came from my heart. It was who I was.

So, having pride in all we do starts inside of us: how we feel about ourselves, our standard. When we take pride in our job it's contagious, it sets a great example. It not only makes us do a better job, but it helps others around us to do a better job.

How do you get that feeling inside you so that it's reflected in everything you do? How do you keep the pride alive so that it lasts through good times and bad?

First, showing pride encompasses many of the things we have spoken about in this book. It is commonly understood as doing the job the best you possibly can. You give any assignment extreme care and attention to every detail. You complete the task and do it in a way that it cannot be done any better—and people want to emulate the way you accomplished the task. You will earn a reputation for that eventually.

Second, you must live it, it must be 100% real. You will never be able to fake it. In order for it to become automatic, it must be you and you must feel good about yourself. It is very hard to do something well over the long term if you don't feel good about yourself.

Ask yourself about the last job, or assignment, or emergency call you went on. Did you give them an experience that showed true pride in what you are and what you're doing? Did it transcend their expectations and make a connection? If not, now is the time to begin to prepare for your next assignment and from here on, doing so with pride.

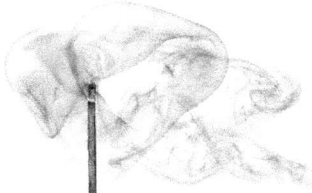

Chapter 32
RUMORS

"A lie can run around the world before
the truth has got its boots on."
— Aphorism

It's widely known that firefighters face incredible dangers in their line of work: fires, hazardous materials, raging storms, and medical emergencies to name a few. Tremendous courage is needed to face those things. However, perhaps the most dangerous things firefighters face are words. Specifically, rumors. Rumors are responsible for damaging and destroying more careers than any other danger.

Is that just hyperbole? We've already talked a little about not getting involved with rumors, but let's get into more detail about their significant dangers, why you shouldn't spread them or let others do so, and why—sometimes—you should pay attention to them.

The harm done by speech can be the most damaging harm done to a person in or out of the fire service. This is very clearly highlighted in the bullying prevalent on the internet—specifically how it damages our young people in America. The old adage "sticks and stones may break my bones, but words will never hurt me," in this case is devastatingly wrong. Reputations can be ruined by rumors, careers lost, they can affect promotions and the attitudes of others toward the individual damaged by the rumor.

I'd personally rather have my leg broken than any of those other results. At least my leg would heal—my career and reputation might not. As has been said, there is no forgiveness for disparaging speech.

A Hasidic tale vividly illustrates the danger of improper speech. It tells this story:

A man went about his community telling malicious lies about the rabbi. Later, he realized the wrong he had done, and began to feel remorse. He went to the rabbi and begged his forgiveness, saying he would do anything he could to make amends. The rabbi told the man, "Take a feather pillow, cut it open, and scatter the feathers to the winds." The man thought this was a strange request, but it was a simple enough task, and he did it gladly. When he returned to tell the rabbi that he had done it, the rabbi said, "Now, go and gather the feathers. Because you can no more make amends for the damage your words have done than you can recollect the feathers."[1] Take the rabbi out of the story and replace it with anyone you have spoken disparagingly about.

> Our words have been compared to an arrow: once the words are released, they cannot be recalled, the harm they do cannot be stopped, and the harm they do cannot always be predicted. Words, like arrows, often go astray.[2] Perhaps you have understood this comparison when you've said something and as soon as the words came out of your mouth, you wished you could take them back. Our words affect every relationship we have in life.

The Talmud says that the tongue is an instrument so dangerous that it must be kept hidden from view, behind two protective walls (the mouth and teeth) to prevent its misuse.[3] The Bible warns against the massive destruction the tongue can cause.[4] The Quran is filled with many verses speaking of the evils of spreading rumors and talking negatively about someone.[5] Buddhists are taught to abstain from false speech, slanderous speech, harsh speech, and idle speech.[6] And Confucius taught about the ills of false speech.[7] How is it possible that all these widely differing religions all say the same thing about false speech and its destructive power?

It is due to the overwhelming power that rumors have to destroy. These religions started in antiquity when people lived in tribal groups. Full reliance upon each other and a family-type bond was necessary for the survival of a group as a whole. If destructive speech were allowed to run through the group uncontrolled, it would lead to its destruction and the destruction of the families.

In the fire service, we too are a family—a brother and sisterhood, and to excel for those we serve and for each other, it's paramount that we avoid disparaging speech. We can and will destroy each other if we let rumors go on. How can a person trust another in the department, or view them as a family member, or want to help them thrive and succeed when that person has spread lies about them that consequently affected their reputation, or a promotion, or even their job? Rumors destroy the bonds of our family and destroy people individually.

When an individual feels the sting of a rumor they retreat into a defensive position, isolated, terrified to trust anyone. Eventually it affects their job performance. If enough people keep calling someone lazy over enough time, it can become a self-fulfilling prophecy.

No human being should be treated like this. When an employee feels like they have to walk on eggshells for fear of making a mistake and their actions become the talk of the department, they will never feel free to reach out and thrive. Others have destroyed that person's potential. It's a wolf pack mentality that does that to others, picking on the new one, the weak one, someone who doesn't fit *their* norm, or someone who already isn't liked because of other rumors.

As a leader, you should never ever engage in starting or spreading rumors. An officer who does or who allows it, shows his weakness, ignorance, and lack of regard for the firefighter, the department, and those they serve. They are a detriment to the fire service. Your job is to lead, to build up others and the department. Rumors do nothing but destroy. Those you spread rumors to will lose trust in you because they'll know that if you're willing to spread rumors about others, you'll do the same to them. If you are there when others are doing so, stop them immediately. Some officers don't because they believe it gives them control: it makes weak people feel good about themselves. When you do not stop the rumors, you are adding to them. You are giving tacit approval that the rumor is true. Your duplicity teaches those you are in charge of that this behavior is acceptable. That is unworthy of the fire service.

But, pay attention a little to the rumors you hear about yourself. Why? There's an old joke, "I like rumors! I find out so much about me that I didn't even know." Sarcastic, but there is a ring of truth to it. Listen to what is being said and evaluate yourself against that criticism. Is there any truth to it? Sometimes people see us better than we see ourselves. In this case you can take a negative and make a positive out of it.

But I also believe in addressing those you hear a rumor from, to get to the source to stop it. Even if they refuse to give the source, you should always tell those saying things about you that it is wrong and you want it to stop. Limit the damage as much as you can. When people know that you will confront them if they spread a rumor, they are less likely to participate in the future. Remember, people who spread rumors are cowards, cowards never like to be approached or called out.

In summary, understand that disparaging speech can be completely devastating to the victim and their career and it's a significant detriment to the Fire Department and those we serve. Never participate in rumors and shut them down when and where you can. If you happen to hear something about yourself, do a deep self-examination to see if there is any truth to it, and if so, fix that flaw.

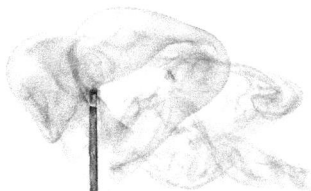

Chapter 33
SHOW THEM YOU CARE

Good leadership can dictate the success or even the survival of a business, a family, a country, a fire department, a military, and just about any other organization. There are books, courses, classes, and seminars devoted to teaching organizations and individuals to lead effectively. But there also are many competing ideas as to how a leader should perform or think. It may be up to you to decide which is the best course for you.

Genuinely caring for those you lead—and letting them see that—is one such area, and I believe this is perhaps the most important principle a leader can possess and practice. In fact, when I decided to write a book, this was the main principle I was focused on. This is the principle I believe makes great leaders great and is missing in officers that can't lead. This is the principle I believe, that helps build the deepest trust and confidence in your leadership.

Not everyone agrees with me on this thought. Many say the opposite and that philosophy has successfully gotten a foothold in the leadership world. One night, I was watching *Star Trek Voyager* season 1, episode 2. Morale is low when the episode opens. The captain is writing in her log about something being wrong with the crew and her feeling not quite sure what to do to make it better.

She then states, "In the academy we were taught not to get too close to those you lead."[1] I was stunned because that is exactly what I had been writing about. In my opinion, it's been misunderstood, misapplied, and therefore has hurt people's ability to lead in the most effective way. If you had to pick one quality as the single greatest quality to true great leadership, what would it be? Stop for a moment and get it into your mind.

Did you think of courage, organizational ability, or intelligence? I would say the greatest quality a person can have is love: love for what you do, love for those you respond to, and, above all, love for those you lead. And if you love those you lead and care about them, then show it.

What do you think of when you think of leadership? Some people see a position of authority where they are in charge and they tell others what to do. Leadership is defined by Peter Northouse as "the art of motivating a group of people to act toward achieving a common goal."[2] But leadership ultimately is a form of service. You are serving your bosses by getting the goal they've set before you accomplished by those you lead. In addition, you are serving those you lead by motivating them and inspiring them to accomplish the goal.

The way you go about this service will determine how good or great and how effective you are as a leader. A factor that will play a role in every interaction you have with those you lead is how much you care about them. It's imperative to truly care about those you lead in order to be a *great* leader. Now this certainly can be a controversial thought. Many see caring about those you lead as a possible detriment or even a weakness.

There is no doubt that goals can be accomplished by those who don't care about those they have authority over. Many subscribe to the notion that "I don't need to be friends or being friends will make it difficult to discipline." I knew a Chief who felt that way. When he was promoted to the chief position, he took the opportunity to visit each crew in each station on each shift. Although the department knew him already, it was an opportunity to say to each one, "Okay, now I am in charge, and this is how I am going to lead." I think that is a great idea. However, he used this excellent privilege to do the wrong thing. His speech basically consisted of, "I am not here to be your friend, I don't need any new friends, we have a job to do."

Now could he be BFF with everyone? Obviously not. But why on Earth would someone go around and tell all those who risk their lives for the public and work hard to make the department look good that, essentially, he really doesn't care about any of them? The effect his words had on that department was heavy. People were actually shocked that he said that. A common retort after his visit was, "Well if he doesn't care about us, why should I care?"

An officer spoke with the new chief and explained the incredible fall-out that was occurring because of his words. The chief stopped saying those things, but his actions continued to say the same things loud and clear. The department suffered from his ineffective leadership and the residents of the city bore the worst of it as the excellent customer service that had been built up for years began eroding quickly.

If you understand that leadership is really a form of serving others then I ask you again, what type of service would you want if you hired someone to do a job for you or your mother or your child? We would all want the person who truly cares about the job they are doing and the customer they are serving.

The customer is always right. You've heard that before. It's the high mark that many businesses try to reach for their customer service and is probably one of the most well-known and overused axioms in the business world. But if it is so well known, why isn't the principle automatic in business? Perhaps it is not the high mark of true customer service. To highlight this point, Simon Sinek, in his book *Start With Why*, relays the following two business examples:

Example number one: "Herb Kelleher, the head of Southwest Airlines for 20 years, was considered a heretic for positing the notion that it is a company's responsibility to look after the employees first. Happy employees ensure happy customers, he said. And happy customers ensure happy shareholders—in that order."[3] So Mr. Kelleher here is saying to focus first on the employee's happiness and they will then make the customer happy.

Do the examples of public companies correspond with the fire service? Absolutely, we are a public service entity. For us, our customers are our patients, the residents of the cities we serve, and our shareholders can be compared to the city's elected officials. Does Mr. Kelleher's philosophy on employee's work? Well, Mr. Sinek continues, "After September 11, there were customers who sent checks to Southwest Airlines to show their support. One note that accompanied a check for $1,000 read, 'You've been so good to me over the years, in these hard times I wanted to say thank you by helping you out.'"[4]

Can you imagine customers of an airline sending money to help them out? That is amazing, but what about the shareholders? Did they suffer because of Mr. Kelleher's care for those he led? Well, Southwest is the most profitable airline in history. There has never been a year when they didn't turn a profit, including after September 11, and during the oil crises of the 1970s and early 2000s.[5] Ultimately, he showed that happy customers are the result when employees are happy. And employees are happy because they know that their leader genuinely cares about them.

Our second example was also highlighted by Mr. Sinek, in reference to one of the most successful businesses in the world. "Walmart did not have a lock on cheap prices and cheap prices are not what made it so beloved and ultimately so successful. For Sam Walton, there was something else, Walton believed in people. He believed that if he looked after people, people would look after him. The more Walmart could give to employees, customers, and the community, the more the employees, customers, and community would give back to Walmart."[6] Under Mr. Walton, Walmart grew to be one of the most successful companies in the world, thriving in different countries and diverse cultures.

JPMorgan Chase is another of the most successful companies in the world. They have over 240,000 employees and operate in over 60 countries.

Jamie Dimon is the CEO of the company and built it into what it is today. He treats all his employees the same, from a CFO to the person who cleans the bathroom, with love, kindness and dignity. When he speaks about his employees you can genuinely tell he cares very much for them. He understands that they are the part of the business that makes the company great.[7] An interesting note: Chase was one of the only banks not to need a bailout during the 2008 financial crises.[8]

In the world of sports, football honors one of its greatest coaches ever by naming its greatest trophy after him, Vince Lombardi. *Inc.* magazine, when talking about love in leadership, remarked, "But another unlikely icon from the past—legendary Green Bay Packers head coach Vince Lombardi—didn't mince words in defining how he led with love. He said: 'I don't necessarily have to like my players and associates, but as their leader, I must love them. Love is loyalty, love is teamwork, love respects the dignity of the individual. This is the strength of any organization. As a leader, loving others is not a feeling; it's expressed as an "action verb." It is *love* that shows up in meeting the needs of others to get results, clearing obstacles from people's paths, and empowering others to succeed and grow as workers and human beings. It has intrinsic value for both leader and employee."[9]

Being in the military and fighting in a war are among the harshest things a person can go through. You have to be extremely tough, both mentally and physically, and this is especially true of those that lead. But does love fit into this environment? Navy SEAL Lieutenant Commander Jocko Willinks spoke about the difficulty he faced when handling the emotions of the team, after one of their brother SEALs had been killed in an operation. When the body of the slain SEAL was being brought in, he knew they couldn't afford to be distracted, as there was still danger to be handled before they could grieve.

He said, "I told the task unit that we would do the only thing we could do, the only thing we should do...put on our gear, lock and load our weapons, and go back to work. Do our duty. The men understood this seemingly harsh course of action. And that is exactly what we did. Everyone accepted this direction from me, not because of my rank or my position, but because they knew something fundamental about me that despite my hardened demeanor and my measured emotions, *they knew I cared about them more than anything else in the world.* This feeling came to me instinctively."[10]

We can go on and on, but I believe the point is made. Truly caring about those you lead and letting them see it is what the great leaders have done in every field.

WHY IS CARING FOR OTHERS SO SUCCESSFUL?

For people to be led to their fullest capacity, they *must trust* the one leading them. That is the key. Even if they are not 100% sure that you as their leader really know what you are doing at every moment, if they trust you, they will follow your lead. Why? Because when they know you care, they know that your directives truly take them into consideration. They don't have to second guess that. If they don't feel that you truly care about them, they will never be certain that your directives take their life and health into consideration.

For example, we know that nothing is black and white about the fire service. We cannot possibly train for every scenario we will be presented with. We cannot possibly have a plan ready to follow in every situation. Sometimes we make guesses based on our knowledge and experience. It is especially at those times that you need those you lead to trust you implicitly and follow your lead.

There is another important aspect to caring about those you lead. As we discuss throughout this book, you are the one training the next leaders. Whether you like it or not, they will pick up things from you and learn from you because you, at this moment, are the single greatest influence on them. Do not fail the residents we serve, the men and women we work side by side with, and those you have been entrusted to lead. What an incredible gift it is to give back to all those who gave you the privilege to serve them.

WHAT IS *NOT* CARING ABOUT OTHERS?

Showing those you lead that you care does not mean that you are best friends or even good friends with everyone. It simply means you genuinely care about them; you genuinely care about what affects them, how they perform, and their career trajectory. When you care this way, your actions will show it.

WHAT *IS* CARING ABOUT OTHERS?

Take some interest in the person. Earlier, in chapter 9, it was brought out how you should meet with every subordinate very quickly when you first start working together and set follow-up meetings. These provide an excellent time to find out more about each person. Incidentally, they are also excellent times to show your care for them. When they see you listening intently to their concerns, when they hear you ask about their goals and how you can assist them in achieving those goals and teaching them to lead, they will sense your desire to help them and that you care. Working together each day gives many more opportunities to get to know them and many more opportunities for your care to shine through.

DON'T EVER FAKE CARING

Caution: don't ever pretend to care about others. This would be worse than not caring at all. If you pretend to care, a subordinate will see that. They may initially believe it, but people are not stupid and will see through your farce. When they do, they will never be able to trust you or fully believe you in anything. In addition, there is a deeper emotional pain inflicted upon an individual who, at one time, thought you really cared but then realized you were faking it.

CARING WILL BE RECIPROCAL AND HAVE BENEFITS

The great thing about caring for others is that it always comes back to you, many times over, and in many different ways. You will endear yourself to those you lead and they, in turn, will do anything you ask them to do because they trust you. In the quote from Navy SEAL Commander Jocko Willink earlier, he discussed the pain he was going through after the death of a fellow SEAL and how difficult it was to tell the men that they had to go right back out there and fight. Not only had these men just heard that a fellow SEAL, a brother and good friend was killed, but now the chance of their own death became much more real as they were heading right back to the line. Yet despite such horrific circumstances, they picked up and went back to work because they knew that their leader cared about them.

Care about your team and they will have your back and want the best for you. They will help you to succeed in your endeavors and make you shine. It will draw your team in very close. I will never forget the love and care that was shown to me by some of the men and women I've worked with and got to know very well. The benefits are limitless.

BE AT LEAST A LITTLE VULNERABLE

Sometimes, when people hear the word "vulnerable," they think it means they have to cry. What it actually means is to put yourself out there, let people get to know who you are: your thoughts, feelings, and challenges. Of course, that doesn't mean you have to let everything out, but enough for people to know who you are and what you feel. This has always been a challenge for me, but I've realized that we don't trust people who we don't know anything about. Do not be afraid to put yourself out there in a place where you may make mistakes. People draw close to those who courageously and humbly handle situations and make mistakes.

Brené Brown, PhD and author of *Rising Strong* stated during an interview in 2013, "A reporter, told me that after reading *The Gifts of Imperfection* and *Daring Greatly*, he wanted to start working on his own issues related to vulnerability, courage, and authenticity, he said it sounds like it can be a long road, can you give me the upside?" Her answer was this, "I believe that vulnerability, the willingness to show up and be seen with no guarantee of outcome is the only path to more love, belonging, and joy." "And the downside?" he asked. "You are going to stumble, fall, and get your ass kicked," was her reply.[11]

It is well worth it. There is nothing greater on this Earth than the gift of love. Give it out generously. Care about each and every one you lead and let them see it. Whatever difficulties arise from it will pale in comparison to all the benefits and blessings that come from giving it.

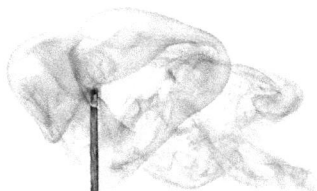

Chapter 34
THE CREW, THE SHIFT, THE DEPARTMENT—THE TEAM

What do you think of when you hear the word team? Perhaps sports? Maybe football or basketball where a bunch of men and women compete against another bunch of men and women on the other side?

What about the team that rose to the challenge to put a man on the moon? The team that created the internet? The teams that have saved millions of lives coming up with the COVID-19 vaccine? All the smaller teams that acted as one coordinated team to save the world from the Nazi aggression?

An individual can accomplish a lot when they put their mind to it. But a team of individuals can accomplish almost anything. The previous paragraph offers just a sample of the great feats that have been accomplished by teams of people. What about you, are you part of a team? You are part of a team that each day is ready and prepared to do battle with the woes of the world that cause pain, destruction, and death. Your team works in coordination with a larger team, your fire department, made up of all the smaller crews in that department. And your department is a team in a much larger team called the fire service. You are an integral part of this. Your team, or crew, is a link in the chain called the fire service. And you are the leader of that team, the coach.

The military is another excellent example of this. We are protected by the greatest military to ever grace our planet, and it's made up of pieces, teams. Small teams or crews that make up a group, groups that make up a branch, and branches that make up the military. Each is like a small spoke in a large wheel.

When you see the larger picture of the whole fire service, it should make you feel proud to be a spoke in this great wheel that is there to help and protect the people who live in and visit this country. Some may feel their part is insignificant, but the opposite of that is true, for if we lose too many spokes, the wheel will not operate.

Your actions, your leadership, your example, can serve to strengthen all the spokes in the wheel called the fire service, but it starts with your crew, your team. You must serve as the coach, helping the team to come together and helping them excel at what they do. You must also protect your team. All too often I have watched as officers pick apart members of their own team, mock them, allow others to mock them, or try to destroy them. If you want to lead the team you must not only train and teach them, but also protect them, so that they can grow and thrive. They will see your protection and they will want to work harder to be a member of the team and make the team thrive. You can't leave them out there to be picked off by others in the department who roam around trying to find someone not protected.

You must protect and defend the unity, the team spirit.

I love tigers and lions, always have. I was recently watching some YouTube videos about lions when something dawned on me. A lion was stalking a massive group of water buffalo. She sprang into action, chasing the herd, and off they ran as fast as they could. The lion chased and chased and before long, a young water buffalo fell behind.

The lion caught the young water buffalo and it was dinner time.

The rest of the herd was gone. I understand that is how lions get food to eat, but there was something that just didn't sit right with me. I started looking at the situation from my standpoint. What if I was with my family, friends, and loved ones and a lion began to stalk us. I would do everything I could to fight off that lion. I would never run and leave my smaller kids behind. All of us would jump in and fight the lion to protect our children, even knowing that we were probably doomed.

But something made the water buffalo situation even more interesting. According to *National Geographic*, water buffalo stand 5 to 6.2 feet tall at the shoulder and are formidable mammals.[1] Males have enormous backward-curving, crescent-shaped horns stretching close to 5 feet long with deep ridges on their surface. Females are smaller in size and weight, but they also have horns, although they are proportionately smaller. They weigh between 1,500 to 2,500 pounds.[2] Their horns are sharp enough and powerful enough to pierce a lion through and flip them 10 feet in the air. There are hundreds of them in a herd, and such was the size of the herd that I'd watched the lion chase and scatter.

Lions, on the other hand, are much smaller, at 4 feet or less.[3] The weight for adult males is about between 330 and 550 pounds and females between 265 and 395pounds.[4] So, I began to wonder what the scene would have looked like if, as the lion approached the herd, instead of running away— every man for himself as we used to say as kids—the water buffalos had turned toward the lion and ran to destroy it.

The lion would not even stand a chance. In fact, a whole pride of lions could not stand up against this herd. The water buffalo would be the kings of the jungle. The lions would be terrified of them if they only acted together instead of sacrificing one, not for the greater good of the herd but for the greater good of oneself. You don't have to be a water buffalo, you can be a lion, but you must stand out in front of any danger that tries to disrupt the unity of the team. You cannot allow someone, either from outside the team or inside, to stalk, disperse and destroy one member of the team.

You have to see the danger approaching and stand out there and fight it like you would if it was your child. Then you will be successful, and in addition to being successful, you will win the loyalty and trust and love of those that you serve, those you lead, and you create a tighter, stronger team. Others too will follow your example, not only refusing to behave that way themselves but shutting it down when they see it. And if your department has developed a culture of leadership, it will be as if the whole herd turned on the lion that tried to disrupt the peace in the department, not only will it not stand but they will be afraid to try it in the future.

Everyone wants to be a part of a team. The stronger and tighter that team, the more that others will not only desire to be a part of it but will fight to keep it cohesive. You are the one to make those steps move forward. Others like you will join with you and the team will grow—the crew, the shift, the department, the fire service. You are a spoke in the wheel of this world. Set the right example and help others to do the same.

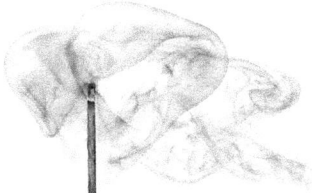

Chapter 35
WHEN THE GOING GETS TOUGH

Being a great leader means not only that you know how to lead, but also that you can handle the tough times. We speak throughout this book about some of the negatives in the fire service and how to handle them, what steps to take to mitigate issues and turn them around. Personality conflicts, rumors, working under terrible officers, being treated unfairly, being discriminated against, politics on the job, terrible assignments, favoritism, being passed over for promotion, and other negatives like these can have a long-term detrimental effect.

It's important to take the steps to try and fix the issues or even prevent some of them from happening. But you will go through difficult times in the course of your career, despite your positive actions. Sometimes the negatives last for months or even years. The longer the problem lasts, the greater effect it can have on us mentally. It can even cause us to want to give up and quit the fire service. How do we get through those times?

Earlier, I spoke about how I felt like I had won the lottery when I was finally hired, and how much I loved the job. However, in a few short years I was frustrated, upset, and completely unhappy at work. I explained that I had been listening to the negative people at the table and that it was affecting me. But there was more to it. I had been battling rumors, personality difficulties (partly my fault), discriminatory actions, and I was working for an officer I just could not stand being around for a single minute, let alone 24 hours at a time.

I was so miserable and didn't know what to do. I started to bring my Walkman in (yes, Walkman. I am that old) and would listen to things during the day to get my mind off work. My officer told me if I brought the Walkman in any more, he would write me up. Although I wasn't breaking any rules, he claimed I might miss an alarm. I was young, not in good control of my emotions, frustrated beyond explanation, and angry—at a boiling point. I was reading a book by Steven Covey, *The 7 Habits of Highly Effective People*, when I came across the concept that no one can make you feel anything, the only person that can make you feel something is yourself.

In other words, the officer wasn't making me upset, I was *allowing* his actions to upset me. I felt like a weight was lifted off of my chest. Instantly, I felt free and light and happy. I went back to work and began to implement that thought. With my new freedom of thought, I examined everything that was going on in the department and why I was upset after starting out so happy. I was able to see how I added to the problems, how I allowed things to affect me, and I instantly went about changing those things. My love for the job began to come back. Of course, it wasn't an overnight fix. I didn't walk in the next day and love the job and love working with my officer, but it was a significant change and—even better—I had a plan, a goal, and a map to make progress. That made all the difference. Day by day, step by step, it got better and better. It wouldn't be the only time I would face mental challenges, where my joy was sapped, but I would come back to this formula again each time, and each time it worked.

More than likely, you will find that your most difficult times in life will involve your mind and/or heart. This is where the control you have developed over your mind and heart and consciousness comes to play. A 14-year-old Anne Frank put it this way in her diary, "Honestly, things are only as bad as you make them."[1]

She certainly was in a most horrific circumstance in life, yet she began to understand that the way she thought about it was going to make a difference in her life. If she thought positively enough, would the Nazis have left and would she and her family be free to start their life over? No, absolutely not. However, as we brought out before, positive thinking opens up opportunities, helps you see the bigger picture, and helps you to not despair. It's a symbiotic relationship.

In fact, although tough times can destroy you, they can also be the greatest growth opportunities in your life. American Supreme Court Justice John Roberts, in a commencement speech to Cardigan Mountain School stated, "From time to time in the years to come, I hope you will be treated unfairly, so that you will come to know the value of justice. I hope that you will suffer betrayal because that will teach you the importance of loyalty. Sorry to say, but I hope you will be lonely from time to time so that you don't take friends for granted. I wish you bad luck, again, from time to time so that you will be conscious of the role of chance in life and understand that your success is not completely deserved either. And when you lose, as you will from time to time, I hope every now and then your opponent will gloat over your failure. It is a way for you to understand sportsmanship. I hope you'll be ignored so you know the importance of listening to others, and I hope you will have just enough pain to learn compassion. Whether I wish these things or not, they're going to happen. And whether you benefit from them or not will depend upon your ability to see the message in your misfortunes."[2]

That is the point! When the going gets tough do you see the message? Do you self-examine and see where you can make changes, where you can work harder, and most importantly, what you can learn that will not only help you through the difficulty but will also make you a better person? A better leader? Keep a clear positive view in front of you. Dutch philosopher Baruch Spinoza said, "Emotion, which is suffering, ceases to be suffering as soon as we form a clear and precise picture of it."[3]

When the going gets tough, work hard through it, but form a clear mental picture of what is happening. Stay positive, do not let the actions of others affect your outlook on the job or on your life. Understand it may take a different approach to work through it. Learn from the experience. *Nothing can break you but you yourself.* Do we have examples of this working in real life? I would venture to guess that if you asked every person who has accomplished much in life, they will tell you of the extremely difficult times they went through and how they processed through them, what they learned, and how they used what they learned on the next trial.

Not everyone that goes through difficult times is successful, not everyone learns something profound from their tough experiences. But that is your goal. If as a leader you can learn profound things about yourself, your work ethic, your treatment of others during these most difficult times, then you will truly be the diamond in the rough: the strongest substance on Earth and also the most beautiful. Victor Frankl, a psychiatrist and a Nazi holocaust survivor, wrote about his experience in the concentration camps, how he was able to survive a hell that most people in the same situation did not, and what he learned.

He stated, "Everything can be taken from a man but one thing: the last of the human freedoms—to choose one's attitude in any given set of circumstances, to choose one's own way."[4] Once individuals in the camps came to understand this, that they could control how they personally thought of the situation, they had a much greater chance of survival and ability to deal with their current circumstances with dignity.[5]

Reading the experiences of others who overcame difficult challenges can help us mentally deal with our own. *Man's Search for Meaning*, by Victor Frankl, *The Gulag Archipelago,* by Aleksandr Solzhenitsyn, and *In Order to Live,* by Yeonmi Park, are books that have given me great guidance. They offer testimony to the thoughts espoused here and they will help you focus your thoughts and attitudes when the going gets tough. They will show you and teach you, that you, too, have the capability, the strength, and the ability to not only go through these tough times but also grow through them. Nietzsche put it this way, "He who has a why to live can bear with almost any how."[6]

Learn your why and do not forget it. You have a job where you affect the lives of those you help, their lives and the lives of all those who love them. You protect the citizens' property and the property of the city. You mitigate the worst of what man does to man, doing your best to alleviate the pain and suffering. And now you are reaching out for or are in a position of leadership, leading men and women in this great work. You are an asset to your department, to your city and to humanity. You will face many difficulties, that is a certainty. But you have the ability to get through them, to not only go through them by to grow through them. Your example will shine for others to follow.

PART 6

INSPIRING THE NEXT GENERATION

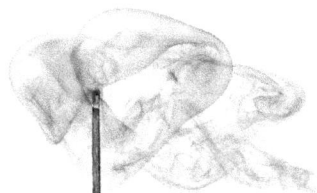

Chapter 36
ASPIRE TO INSPIRE

The deep gray sky attested to the time of the year. Bitterly cold air tried its hardest to break into the house through the old drafty windows. It was the weekend! No school! Time for football! We would meet in the streets of Brooklyn for big battles all day. It would be grueling hours of pounding, running, hitting, catching, and tackling. But before we headed into the frost to clash like warriors, it was time for some inspiration. For me, it was time to watch Walter Payton.

My brother slid the video cassette into the VCR and the deep voiced NFL announcer began his prologue, "In the annals of football legends, one man stands above them all." Boom, the music strikes a fast-paced upbeat anthem and there he is, the man, running around tacklers, over tacklers, and through tacklers. I felt the energy and excitement building in my blood. The longer it went on, the more amped up I became. I felt like I could run through a brick wall.

There I was, a young teenager, feeling like a professional football player. Out the door I go, ready to destroy the competition. Watching Walter Payton could always make me take my game to the next level. He inspired me in more ways than just playing football and more than a paragraph can explain. I wanted to be a football player more than anything when I got older. That didn't pan out, as a lack of college funds killed the dream. But Walter Payton still inspires me today—not to run over people anymore, but to be the best I can be, to compete without letup and endure any and all trials.

What inspires you? What inspired you to get where you are? What is inspiring you to go to the next level? Who has inspired you to become better? Can you do that for someone else? Can you do that for those you lead? The answer is "yes." The better you inspire, the more of a response you will get from those that you lead. But inspiring takes more than just a video, a speech, or an action. It takes a whole series of things. It takes time. It takes you doing the right things, living the right principles, setting the example, being disciplined, and getting personal results.

In his book, *Start With Why*, author Simon Sinek discusses the style of the first two CEOs of Microsoft, Bill Gates and Steve Ballmer. He states that while addressing shareholders, Mr. Ballmer is very energetic, runs around the stage sweating, and is very motivational. But he further states that that motivation is short lived. On the other hand, he says that Bill Gates is "Shy, awkward, and a social misfit" and says he "does not fit the stereotype of the leader of a multi-billion-dollar corporation."[1] In this juxtaposition, you are led to see a stark difference in leadership between the energetic, motivating, charismatic style of Mr. Ballmer and the shy, slow, methodical, socially inept Mr. Gates. Perhaps this already brings to mind who is a better leader.

Despite his social failings, Mr. Gates has far greater long-term effect on those he speaks to and those he leads than does Mr. Ballmer. Mr. Sinek writes, "When Bill Gates speaks, however, people listen to him with bated breath. They hang on his every word. When Gates speaks, he doesn't rally a room, he inspires it. Those who hear him take what he says and carry his words with them for weeks, months, or years. Bill Gates doesn't have energy, but Bill Gates inspires. Energy motivates but charisma inspires."[2]

While I believe what Mr. Sinek is saying here, I want to take it one step further. I believe what lends credence and power to Mr. Gates's words and what makes them inspiring is that they are backed up by his accomplishments, the results of his efforts. What you know of Bill Gates is that he is a brilliant human being, he started a company very young, dropping out of Harvard, he ran the company and built it into an international business, one of the most dominant and powerful and influential in the world.

To add to that, he eventually stepped aside as CEO but continues to influence the company as a board member, so that he can focus more time on philanthropy, of which he is one of the most generous people in human history. By way of his and his ex-wife's foundation, he is helping to attack issues that devastate people all over the globe. He has already been responsible for saving countless lives.

Mr. Ballmer on the other hand—and I mean absolutely no disrespect toward him because running a huge company is a difficult job—presents a different story. What is generally known of him is this: He did not build a company but instead inherited one of the most powerful companies on Earth and during his 14-year tenure as CEO, the company floundered. Microsoft fell behind its competitors, several huge purchases failed, and its stock price stagnated. On the day he announced his retirement the stock soared 7%.[3]

So I ask you, if Bill Gates's résumé was the same as Steve Ballmer's, would people wait on his words with bated breath? Would they be inspired by a "shy, awkward, social misfit" and carry his words for weeks and months and years? The obvious answer is no. So, the weight of his words comes from his accomplishments. His accomplishments come from years and years of hard work, actions, discipline, doing the right thing, and setting the right example.

The title of this chapter is "Aspire to Inspire." How do we do that? I wanted to set the foundation for you. First, a desire to inspire others must be a core part of our belief. It is something we have spoken about in this book several times. You must feel and know that it is imperative that you lead, train, and inspire the next generation to take the lead.

In his book, *Extreme Ownership, How Navy SEALs Lead and Win*, Jocko Willink plainly puts it this way, "The goal of all leaders should be to work themselves out of a job. This means leaders must be heavily engaged in training and mentoring their junior leaders to prepare them to step up and assume greater responsibilities. When mentored and coached properly, the junior leader can eventually replace the senior leader, allowing the senior leader to move on to the next level of leadership."[4] If you feel this way, then each day it will be a part of your performance, to help others.

We must recognize that the first step starts with us individually. We have to be a good leader. We have to set the right example always. Think of a terrible leader that you may know, in or out of the department. Would he be able to inspire you? Now, he may give you examples of what not to be like, but inspiration must come with good results. Why? Because accomplishing anything of significance in our life comes with difficulties. The greater the significance the greater the difficulties. That is why there is a difference between motivation and inspiration. *Motivation can help you through a task, inspiration can help you through a life.*

When you come up against mountain-sized difficulties, it is inspiration that will help you through them, not motivation. Motivation will get you to the first crest, inspiration will get you over the top. Why? Because inspiration goes to your core, it changes who you are at the innermost level. So, as you continue to work hard, first on yourself—doing and setting the right example, working to become a better person, a better firefighter, and a better leader—you will rack up accomplishments. Those accomplishments will be seen by others, especially those you lead, and your words will have weight, they will inspire, they will "stick with people for weeks and months and years."

Envision what the perfect fire department would look like: Incredible customer service. The residents love the department and constantly show it. The brothers and sisters in the department truly care for each other and work hard as a team to accomplish everything they face with precision and perfection. The firefighters love their job and continue to better themselves and help others. It is a happy, healthy, and productive workday each day. Can you see it? Can you feel it? Impossible?

Each day aspire to inspire others to your perfect vision of your department.

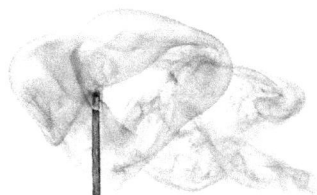

Chapter 37
DEVELOPING YOUR TEAM MEMBERS

An often-overlooked part of leadership is developing your team members. Sometimes it seems like those in charge want to be in charge forever. Yes, it's true. As we have brought out, leadership is about accomplishing a set goal in front of you. However, in the fire service, when you are appointed to a leadership position, one of your objectives should be training the next generation of leaders. If you love the fire service and you are good at what you do, you will already want to train the next leaders so that in your absence, the department continues to excel.

It must be said at this point that not everyone wants to reach out for a position of leadership, and that is perfectly fine. Leading others is an extremely difficult task. As we've discussed, it takes continual self-examination, working on deficiencies, knowing your job well, caring about others you are leading, and helping them to succeed—and all on top of the emergency tasks that constantly come your way. Another consideration is that it is a position that never ends. You can't be the leader today and tomorrow let someone else handle it. Once you are in that position, you are in it, night and day, until you get promoted or retire.

In addition, not everyone is built to be a leader and if someone has no desire for that role, they will not be a good one if forced into that position. Many times, in the fire service as in the business world, people are promoted past their competency level. This ends up being detrimental for them, for those they lead, and for the service. For all the individuals who do not wish to reach out to a position of leadership, it should not be held against them, nor should they be made to feel badly about themselves. It is an excellent opportunity instead, to train them to the maximum level in their position. What fire officer would not want to have a veteran, well-trained, exceptional driver/ engineer on their crew and firefighters who love what they do and are great at it?

So how do we develop our team members?

IT STARTS WITH US

Our first step should be, as with all things, ourselves. Are we working to be a good leader, are we setting an excellent example, do we know our job well, do those we lead know we care about them, are we someone they want to follow? You can't develop others as leaders if you can't lead yourself.

HELP THEM EXCEL IN THEIR ASSIGNMENT

The next step should always be to develop our team members to be able to do their assigned jobs. We want to help them excel in their assignments, feel proud of their accomplishments, work hard to be an integral part of the team, and give the best possible customer service. Whether they want to reach out for leadership or not, their current job assignment is a priority we should focus on. If during that training they express a desire to be a leader one day, or you see what you perceive to be leadership qualities, you can begin to work on developing those things.

BE COMPLETELY OPEN AND HONEST

We want to be completely open and honest when helping them. Don't hold anything back, thinking you need to be better than they are. Don't fear helping them to progress. Yes, they may be your peer one day, maybe even your superior. Don't fear to help them get there. A true leader never hinders progress, instead he seeks and develops it.

Aleksandr Solzhenitsyn put it this way: "Talent is always conscious of its own abundance and does not object to sharing."[1] It could be said then that when someone fears training others, it is because they recognize their own lack of talent or lack of leadership. When you understand we are a team—the whole department—you will want to help the department succeed.

Let me give you something to reason on; sometimes in sports this type of situation plays out. A team has an excellent player and he has been looked to as the leader on the team for some time. After much time has passed and that leader no longer is at the top of their game, the performance of the team drops. Then they draft a young player just full of talent and promise. When they play together, you may hear the two of them referred to as the teacher and the student. In time, the student passes the teacher and becomes the leader on the team. Is it a bad thing? No, the new leader takes the team to new heights—perhaps even a championship. All enjoy the success including the teacher.

DEVELOP A LEADERSHIP CULTURE

When we are developing leaders, we are creating a culture in our Fire Department, and we are ensuring not only the future success of the department but also that it continues to excel. How? *Develop a leadership culture.* Mark Miller, in his book *Leaders Made Here* wrote, "How do you ensure you'll have the needed leaders to fuel your future success? The answer, in short: Build a leadership culture. Let's be clear on terms from the beginning. A leadership culture exists when leaders are routinely and systematically developed and you have a surplus of leaders ready for the next opportunity or challenge."[2]

I believe this is imperative, and not just for future success: it will bear much fruit immediately as well. John Maxwell, who has focused on leadership for many years, explains it this way, "A company's culture is the expression of the values of the people within the organization. It is the sum of the behavior of the people, not a reflection of what you want it to be. People do what people see—and they keep doing it. What people do on an ongoing, habitual basis creates culture."[3]

Having a leadership culture will affect and infect all those who work in the department.

STEPS THAT CAN BE TAKEN

If you are the leader of the department, generally the chief, you probably already meet periodically with your staff and go over pressing issues and the needs of the department. Perhaps on the agenda should be the identification and development of leaders in the department. I recommend discussing this point at least once a month. These individuals, identified and developed, will end up inevitably leading and running the department. All the resources needed to help develop the next leaders can come out of those meetings.

A strong caveat must be discussed here. Many times, leaders tend to look for and try to develop followers instead of real future leaders. Or they may identify and develop their friends and buddies. Everyone in the entire fire service has heard of and probably experienced the good old boy system. As was brought out in chapter 29, a good leader does have to be a good follower, so it does not mean that a good follower would not be someone to develop. They may definitely be a good leader.

However, do not discount someone because of a personality conflict. A diamond starts as carbon, a piece of coal, the thing Santa puts in your stocking if you are bad. Then with heat and pressure and time (development), it becomes the most beautiful and hardest stone on Earth. So that person you don't like, with some time, care (heat), and guidance (pressure) they may become the greatest leader the department has.

I love the way this exercise is referred to in John Maxwell's book, *The Leader's Greatest Return*, "When questioned about his stone masterpiece "David," Michelangelo is alleged to have said that the sculpture already existed within the stone; he simply had to chisel away the rock around it. That is what leaders do. They see the future leader within the person and they help that leader emerge. Maybe that's why professor and bestselling author Brené Brown defined a leader as 'anyone who takes responsibility for finding the potential in people... And who has the courage to develop that potential.'"[4]

And never *ever* let race, religion, sexual orientation, or ethnic background prevent you from identifying and developing someone. No one should ever be discriminated against. A diverse set of viewpoints can add great strength to the department, and let's not forget the fact that we serve a diverse group of Americans.

Identifying and developing men and women in all positions in the department is essential. Those already in leadership positions—lieutenants, captains, and above—are a clear pool to begin with. But it would be exceptional to begin identifying and working with new hires as soon as they get off probation as well. This allows you to start off slowly, evaluating and watching how that individual handles assignments or projects given to them. It also allows the culture of leadership to be instilled right from the beginning of their career.

This gives you the opportunity to work with the FF's officer to help them continue to progress as a leader and to help them to train others as leaders as well. We learn best when we learn something, apply it ourselves, then teach it to someone else. This imprints the lessons and principles into one's heart. It becomes their core.

Setting a leadership culture throughout the department will be reflected outside the department as well, in customer care, customer service, and interactions with other departments both inside and outside your city limits. City leaders will appreciate the atmosphere, residents (customers) will praise the actions of your department, and departments and firefighters from all over will attempt to emulate. The best will try to flock to your department.

Can you imagine, when getting hired, a firefighter hears from others that the department has an excellent leadership culture? Right from the outset, that employee has bought into the fact that the department they are going to work in is exceptional. This sets an exceptional standard for them to work toward right from the beginning of their career and something to be even prouder of. It would be like a wide receiver not only getting drafted into the NFL but being drafted by the team Tom Brady plays for, or a young person not only getting accepted to college but being accepted by Harvard, or a person not only being offered a job in a tech company but being offered a job with Microsoft. Automatically, these individuals know they are a part of something special and that they have to perform at a higher level as well. If you have a leadership culture in your department it will become the standard all other departments will strive for.

Programs can be developed to help implement this culture and develop these leaders based on information found in this book and other books and materials, fire department policies and procedures, and city guidance and ordinances. A department vision can be developed and shared by all, with the example being set by the current leaders. What an incredible start it would be to get input and effort from all the current leaders or even the entire department, depending upon its size, to put a program together that will encompass the culture of the entire department, the entire team. When all feel they had a share or some input, they will be more likely to work to make the implementation a success. Once a department has a culture of leadership, it will be hard to reverse the trend. Leadership will continue to get better and the department will continue to excel.

Developing the next leaders can seem a little scary to some, however if done correctly, all in the department will benefit as well as all those we serve. As a middle management officer, you may not be able to develop such a program on your own, so perhaps you can make the suggestion and help work to facilitate it. Also, you can always do it on a smaller scale, with your immediate crew, in fact that is where you will start.

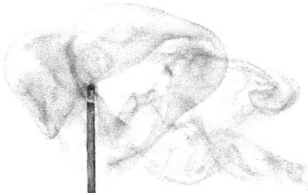

Chapter 38
THE DIFFICULT EMPLOYEE

Working with a difficult employee can be one of the most discouraging and difficult things for a leader. The main reason is because the employee's behavior, to a degree, is out of your control; there is no guarantee you will be able to help the person. Their behavior can add an enormous amount of work for you. In addition, their attitude can be a negative influence on you and other crew members, making your workload more difficult exponentially.

A common tactic is to leave the situation unaddressed. It is certainly easier that way. However, if you do, the difficult employee can start to see you as weak and become emboldened to worsen their behavior. In addition, he or she will affect your attitude and the attitudes of the others on the team, dragging them into negativity or destroying things you had built up. Perhaps you have heard the saying, "if you lie down with dogs you will wake up with fleas." Every interaction we have will have an influence on us. The more difficult the employee, the more influence they will have on the group. So, you need to address and fix the situation.

A second ineffective tactic is to try to create an "us versus them" mentality. An officer may make statements about other firefighters in the department such as, "they don't do this as well as us," "we work hard and they are lazy," "we are better firefighters than they are" and the list goes on. How many times have you heard similar statements about the crew you relieve each day, that they don't check the units out, or clean the station well, or train much?

Leading the employee or the group to feel that they are better employees than the other groups, or shifts, is a very short term, short-sighted, and detrimental fix. I worked with an officer who propagated this attitude and I saw people buy into it, especially developing a wolf pack mentality, where the group was encouraged to attack the job actions of a single person or even a group of people.

I saw the tremendous holes in such a style and the divisiveness it created, and I was shocked by how many supervisors used this style. Supervisors will use this style of leadership out of laziness and ignorance. It's detrimental because it begins to break down the team approach of the crew and the whole department. Individual companies begin to develop animosities toward others. Of course, this leads to them not working together as a team. Negative comments that are made about others come back to them and can lead to animosity. A fire department is a *family*. When there are divisions within the family, the whole family suffers.

On an individual level it can be catastrophic. As a supervisor, your goal is to integrate the individual with the team goals and the team with the department's goals. By using the "us versus them" tactic you are creating a negative, destructive attitude in the employee. This negative and destructive attitude can be especially influential on newer and inexperienced firefighters. The more negative the employee gets, the more he will hate his job and the more others will begin to distance themselves from him. A reputation can develop which can even affect this individual's chances for promotion in the future. It is a vicious cycle that continues to spiral downward.

Now let's look at a few steps that can help.

1 - Meet with your supervisor.

This step depends upon the severity of the situation. Most of the time you will not need to take this step, however in those few significant situations when you do, you want to inform your superior of the situation and your course of action. By this time, you should have documentation, evidence of inappropriate behavior, and evidence of the attempts you have made to correct the situation in the past. Your meeting with your superior is to ensure that you are both on the same page with your future course of action. If your superior does not agree with you, you will have a difficult time using discipline as a tool for corrective action. If they do not see it the way you do, it's an opportunity to get some guidance from them. This will help arrive at a plan that has you both on the same page, which you may need if you have to escalate discipline and it's an excellent teaching moment. I have learned so much from others on how to deal with individuals when I have a disagreement with someone.

2 - Meet with the employee.

Whether you had to meet with your supervisor or not, meet with the employee as soon as possible and include the next three steps in your meeting. Before ending your meeting, ensure you are on the same page, that you both fully understand the problem and the resolution. Set up a follow-up date *before* your meeting ends.

3 - Show them you care.

Chapter 33 is devoted to this subject, and there's an opportunity to apply that information here. Show genuine concern for the employee. By your tone, your facial expression, your body language, and the words that come out of your mouth, let them know you care about them, that you want the best for them. Do not use condescending language or belittle them. Do not raise your voice or disrespect them. Good leaders genuinely care for those they lead. If it is a new assignment, you may not know the person well enough to have developed a relationship or to have found their good qualities. Perhaps they have done things to make your life more difficult, but that is where genuine care for fellow human beings will come into play. Genuine care for fellow firefighters will help you show that you care for them.

4 - Lay out the fire department expectations and your expectations.

Be clear and concise. Let the person know that the behavior will not be tolerated. People genuinely like black and white expectations—they're easy to follow. This is where your excellent example of "always doing the right thing" will make your job easier, a lot easier. If the person being disciplined is able to point out that *you* routinely do not follow department expectations, your words will be viewed as pathetic. You will never win the person's respect, you will never lead them. You will be able to force them with the threat of discipline, but you will never be able to lead them because they will view you as a hypocrite.

Document your conversation, the deficiencies and the expectations. Before I leave any important meeting, especially one where there was a disagreement, I make a point of saying, "Okay, I want to make sure we are on the same page, we both understand each other." Then I go over our conversation point by point. This lets the other person know I heard them and understood them, and gives full clarity of what the problems are.

5 - Don't withhold discipline.

This was one of my big mistakes on my first assignment as a new officer. I thought that talking to the employee repeatedly and not disciplining them would make me a good officer. Instead, it did the opposite. It made me a bad officer and a terrible leader. I was assigned to a crew that had a driver who had about eight years more time in the department than I did. I did not realize until everything exploded that he resented the fact a junior man was his officer.

He kept acting out and I kept talking to him. But the fact that I didn't discipline him only emboldened him to be even more brazen and that made me look very weak to the other crew members. He finally caused a major scene on a medical call, which was detrimental to the patient, and I could not stand by anymore. I talked to him and wrote him up. He exploded and began a systematic effort to cause me problems. It seemed as if he was waiting for that moment to put his plan into action.

He went to the chief with a list of gripes about me, things I had done during the time we'd worked together that he thought were wrong. His goal was to try to get me demoted. He requested to see all of the evaluations I had received since I was hired, so that he could try to find something in them to help his cause. His list of gripes were minor matters, so he was not successful in getting me demoted, however I *was* moved to another shift. Although it was a blessing for me to get away from that driver, my direct superior acted weakly by moving me and the subordinate would use his weakness against him again and again. He paid a terrible price for that mistake.

That was a lesson burned into me. While disciplining someone by writing them up always remained a difficult thing for me to do, I made sure I did it when it was necessary. The interesting thing is, as I became a better leader, I hardly ever had to write someone up again. Discipline takes all forms: for those in middle management writing someone up should be the last step.

6 - Meet regularly.

Depending on the severity of the problem and the extent of it, you may want to meet again with that employee within a week. If the offense was not too severe, you can extend the time, but don't delay a meeting for too long. Meet regularly. Remember when we spoke about evaluations for your subordinate—if you have any input or responsibility in that process, it's advisable to meet quarterly. When dealing with a difficult employee, after your first follow up meeting—again depending upon the severity of the problem—you might want to meet monthly until you feel strongly about the subordinate's progress.

7 - Help them love the job.

Dealing with someone who makes your life more difficult is frustrating and can be upsetting. But keep your goal in mind, it will make things easier. Your goal is not to punish the person, not to kick them out of the fire service or to try and make them pay. *Your goal is to help them love the job again.* (And if they've never loved the job before, your goal is to make it possible.) If they love the job, they will love working, they will love the department, they will love being part of a team. How do you do this? Help them see the beauty in what they do. Get them more involved. People become bitter for so many reasons, but to be truly happy they have to make progress—help them set up goals and make progress toward them. If you do, you may see a miraculous change.

CHANGING ATTITUDES

While dealing with a difficult employ can be one of the most challenging experiences you go through, it can also be one of the most rewarding. Let me give you an example. One day, the chief came to the station to see me. He told me I was getting moved to another station. He told me that the officer I was replacing was not handling firefighting business correctly and the crew was being affected.

On my first day in that new assignment, at 0810 hours I noticed one of my firefighters in the station's hallway talking with a guy who was going off duty. He was still in street clothes and unshaven. Our shift change had happened at 0800 hours. I was confused and figured that someone must have swapped shifts with him—perhaps I had not seen the latest roster. I asked him about it and he said "No, I'm working today," and then went right back to his conversation, like I was not there. I was shocked.

I looked at my watch and asked him if his gear was on the engine. He said, "not yet." I said please get in uniform, shave, get your gear on and check the engine out (our normal duties in the morning). He gave me an irritated look and went upstairs to the locker room. When he emerged over an hour later, I asked him again, is your gear on the engine? He said, "I am going to put it on right now." It was 0945.

After the crew finished checking the engine, I called a meeting.

Normally, I would have called in the individual by himself for a one-on-one meeting, but it was apparent the entire crew was on the same page. It was all of them against me. The previous officer had let them do whatever they wanted, and they wanted things to stay that way. (I was completely shocked when I heard some of the things these guys were doing.) I began the meeting by telling them what the fire department expected of them and what my expectations were. The conversation exploded into yelling—they were furious.

I did not raise my voice, I simply told them that if they were unshaven, not in their proper uniform, or didn't have their gear on the engine by 0800 the next shift and all shifts after, they would be immediately relieved of duty. To say they were angry is putting it lightly. I ate alone, none of them talked to me unless it was necessary. At night, if they were together and I entered the same room, they all walked out. We had a meeting about behavior every single shift for about six weeks. The entire crew met with the chief and demanded that I be moved. The chief said that was not going to happen, and that as long as I was asking them to do their job, he was going to back me. That was a huge help.

The crew tried, every shift, to get me to ask for a transfer. They did everything they could to isolate me. It was miserable. I spent every shift with a crew that was angry at me, conspired against me, and ignored me, except at the meetings, which were very emotionally charged and loud. I had worked with one of the crew members earlier in my career when I was a driver and he was a firefighter and we had become close friends. Several years had gone by since then, but I thought we were still friends. He asked to meet with me privately after several shifts.

He asked me why I was being this way. What was wrong with me? Why did I change?

"I didn't change," I said. "You did. You've lost your way—that's why you're miserable here and hate your job. Just do what I ask, and I promise you will love coming to work."

"I don't think that will ever happen, but we'll see," he said.

"Yes, you will."

One guy who only had about 14 months on the job, routinely lectured me at our meetings, telling me what a good officer would do and how I wasn't one. Right from the beginning I explained to the crew that if they gave it a chance, they would love firefighting again. Another crew member kept pressing the chief until he was finally able to get a transfer out, and that fairly new firefighter who lectured me got moved to another station as a floater, so he didn't work with us much anymore. Meanwhile, I kept moving toward the goal.

We trained constantly, partly because I believe it's necessary to stay great at our job and partly because they really needed the training. I reached out to a neighboring city and was able to secure an empty warehouse. Shift after shift, we went and trained there on all kinds of fire tactics, search and rescue, RIT, smoke out drills, hose drills, entanglement training, and much more. We trained all summer in the Florida heat, it was 100 degrees with the heat index outside and inside the warehouse was even hotter. We drilled for hours every shift. Little by little, the crew began to come around.

They slowly developed some pride in their abilities and in us as a team. We started setting up training for others who were interested and even built training props. Our days began to be fun again, and the crew began to be looked at as top guys in the department. We were a team. Finally, we were eating together, work was a blast, they loved their jobs, and I loved the crew. It was a tremendously rewarding experience. Everything I thought about building a team was put to the test and it was successful.

The story continues. The firefighter who met with me privately early on fell in love with his job again, and he was good at it. He loved training so much that when a position opened for a FTO (field training officer), he put in for it. He asked me what I thought, and I encouraged him to go for it. It was an easy one. He got the position. What an incredible turn around. I felt such pride and excitement. The only issue was that he had to move to another station.

When he left, I made a request for a specific firefighter. I had worked with this firefighter when he was on probation and I thought well of him. However, when he got off probation, another officer had said some negative things about him, saying he was lazy and had a bad attitude. These rumors spread and stuck to him. I felt differently and personally requested him. My boss, the battalion chief, asked me why on Earth would I want this guy. I told him I believed he could excel.

He hesitated. I said, "Give me a chance to change his attitude, I believe I could repeat what I did with the crew."

He was willing and said, "If it doesn't work out, we can move him back."

The firefighter I requested came up and he thrived. What an incredible addition he was to the crew. He loved it. And an interesting thing happened next. That fairly new firefighter who had lectured me on how to be a good officer asked to meet with me when he heard I had requested this other firefighter.

When I met with him, he explained that he wanted to come work with our crew. He wanted to know why I had not requested him. He actually started crying. I was in complete shock and felt bad for him. As a floater, he had floated in, worked with our crew a few times when one of the firefighters was off and he'd loved it. I did not rob him of his dignity.

I said I was sorry that he felt hurt by me. I explained that he had given me such an incredibly hard time during the months before he was moved out of my station. I said, "You went to the chief behind my back several times and tried to get me moved. You tried to make me look bad. Why would I ever have asked to have you on my crew?"

He explained that he had changed. When he worked with us as a floater, he saw what a great time the crew had together, how much they loved the job, and he wanted to be a part of it. I explained that my decision on the requested firefighter was final, but I promised to give him another look if another position opened up.

This crew and the one with my friend before he left to become FTO were by far the two best crews I ever worked with. We absolutely loved working together. To this day, I still keep in touch with all of those guys. I love them like family. The misery I went through for a couple of months was turned into such a success that it became a true highlight in my career. It gave me a tremendous amount of confidence.

It was noticed by the chiefs, and they began sending me other individuals with what they termed "attitude problems." By that time, it was much easier to help these firefighters as the whole crew was on the same page, they loved what we were doing and we were doing a lot. I found that most of the firefighters that came to be with us just wanted to love what they were doing and work next to men and women who were good at what they did and took pride in their work. It was incredible to see the changes in people and to see their pride in what they were doing each day overflow.

The incredible joy and satisfaction I had from the outcome of these events made me feel that being sent to that difficult crew was one of the best things that happened to me in my career. I helped turn these men's careers around and they, in turn, helped me get better. I helped them to become an excellent example to the department. I helped them excel in their abilities and they in turn helped others. And finally, it was a benefit to the men and women they served.

On top of that, the experience bettered me in so many ways. I too became a better firefighter, a better officer, a better leader, and a better man all the way around. It was something that would give me many positives in my life even outside of the department. My wish for anyone who reads this book would be that you, too, have such an amazing experience.

Having a difficult employee on your crew can be exceedingly difficult, but you have the ability to turn it around and make that person excel. Remember that nothing worthwhile comes easy. Keep your eyes focused on what the goal is, work hard and reap wonderful success, it will change your life.

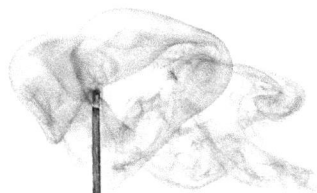

Chapter 39
EMPOWERING NOT MICROMANAGING

If there is anything that kills progress, trust, and leadership skills, it's micromanaging—constantly going over and checking on the actions and progress a person is making on a task they were assigned. Micromanaging is giving too much input, feedback, criticism, and asking too many questions as the task progresses. It is "fixing" things your way when a person completes a task. It's the belief that you would have more success just doing a task yourself, so you dictate to the person *how* to do what you want them to do.

Are you a micromanager? Here is an excellent checklist that Max Freedman, from *Business News Daily*, put together:

- You personally take on work you've assigned to your team.
- You often walk through your office [station] to check what your employees are doing.
- You require that your employees copy you on all emails.
- You frequently ask for updates on tasks and feel a nagging need to know all of your employees' whereabouts and activities during work hours.

- You focus on superfluous details and pedantic errors instead of larger concerns.
- You provide strict rules about how work should be done, instead of giving your employees the freedom to complete tasks as they see fit.
- You ignore your employees' thoughts and opinions when assessing the quality of a project.
- You feel the need to have the final say on all decisions.
- You almost always find yourself unsatisfied with your employees' final work.[1]

Have you ever wondered what goes through a person's mind when you micromanage them? I can tell you it's not, "Oh, I am so lucky to have this person tell me at every little step to make it the way they want it." A person who is micromanaged feels you don't trust them. They might also feel that you think they're too stupid or too incompetent to accomplish the task.

Are you a micromanager? It can be difficult to know for sure because most of the time micromanagers don't know they're doing it. They just believe they know better and need the task done their way—the right way. Perhaps you have been micromanaged yourself.

Examine your relationship with your subordinates to see if you are micromanaging them. Review the list above. Ask a peer officer or your supervisor if you have that tendency. Do not rely solely on your own opinion.

When you're about to assign a task to someone, think about the task before you make the assignment. Make sure the subordinate has the capacity to do those tasks. Build off the strengths of those you lead. Ensure that all the tools the person may need are available to them. Make sure they know and understand that you are available to them for any help or further guidance as they work through the task. Encourage them along the way and butt out.

When the task is completed, refrain from trying to change the results to the way you would do it. Many times, your way may not be the only correct way to do a task. Your way could be the wrong way or an inefficient way. Remember that you are not perfect either.

Of course, this doesn't mean you should let things slide when something is wrong. In fact, it provides an excellent opportunity to help the person examine what went wrong, fix it, and learn from it. It will also help them handle the next task you give them.

In addition, if someone fails at a task, you must ask yourself, whether you failed them somehow. That is where empowerment comes in. Did you give the person thorough instructions? Was the task in a field outside their experience? Were time deadlines impossible to meet?

This is not to say that every time a subordinate fails it's because you failed to do something right, but there are certainly occasions when the person making the assignment could do a better job at helping the subordinate succeed. Remember your ultimate goal is not only to get the task done but to help the subordinate succeed. You want them to finish the task successfully and feel good about it so that they not only learned from it but are ready to reach out and accomplish the next task.

When a shift is over or a task is completed, do those you lead feel empowered, energized, and excited about the next task? Or do they leave dejected, disempowered, and just happy it's over? Do they trust that you are there to help them succeed? Do they feel appreciated for the work they've done? Or do they feel they were treated like a child?

The power is in your hands to break people down or to build them up, to make them feel they've failed or to make them feel successful. I recommend that you not only do a good self-examination on these questions but that you revisit the questions from time to time.

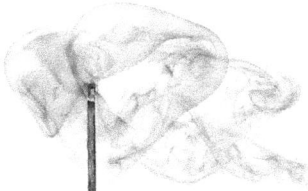

Chapter 40
GIVE BACK

As firefighters, we have one of the greatest jobs in the world. We care for the community and the people we serve. We meticulously care for the stations, engines, and units that the community provides for us. But we can always go a step further. With such responsibility and the ability to handle, mitigate, rectify and correct such major emergencies, can you go further to give back to the community? In this chapter, I've listed some ways you might do that. Becoming involved in many of these actions and programs might need to be approved by your department. You should discuss them with a supervisor before you do anything.

- Offer an outreach program to teenagers in the neighborhood. Show them the beauty of our business. Give them a tour of the department or units. Introduce them to the crew, make them feel like the fire service is a part of the neighborhood they belong too. Help them get into volunteering if your department has a program for that. If not, perhaps another local department has a volunteer program.
- Does your department makeup reflect the community it serves? If not, can you set up a program to help seek out and recruit

members of the community.

- Volunteer to speak in schools, read to children, expose teenagers to the pros of the fire service, and help them get on a track toward service. Shine as an excellent example in the community. Be someone young people can look up to.

- Bring the fire engines, ladder trucks, or medic units to schools for young ones to see. Small children love to try out the hoses and spray water, sit in the front seats, and pretend to talk on the radios. They always enjoy the back of the medic units, playing around with the equipment and watching the crew show them how some of it works. For teenagers, set something up to let them see the challenges of the job and the wonderful benefits to society and themselves. Discuss the challenges they face as young people in society today and how a career in the fire service is an excellent way to give to others.

- Set up a program to check the blood pressure of residents free of charge from time to time, or perhaps administer flu shots. This affords you an excellent opportunity to share information on caring for one's health: blood pressure, diabetes, heart disease, and other health risks.

- Offer the community educational classes on topics like CPR, fire safety, how to use a fire extinguisher, what to do if someone is choking, how to prepare for a hurricane and other emergencies that affect your locale.

- Set up a firefighter appreciation day at a park or another open environment, with units for demonstrations, information, games for the kids, health-care tents etc. The goal is to make a warm relationship with the city you serve. It's an excellent way to draw the community together.

- Reach out and mentor a young person preparing for a career in the fire service with one-on-one discussions and personal attention and perhaps working alongside their parents with guidance, information, and encouragement. Give them help when they are studying at fire or EMT school.

- Set up a program to help parents learn about safety: putting children in car seats correctly, the dangers of space heaters and smoking while in bed.

The number of ways that you can give back to your community is endless. All you need is a desire to do so and input as to what your community might need. Use your imagination. The benefits to the community can be enormous and you will benefit as well. Have a welcoming and warm attitude when others come to the fire station. Set an example for city residents, young people, fellow firefighters, supervisors, city officials, and others. Your department can earn a reputation of caring for the community. What could be better?

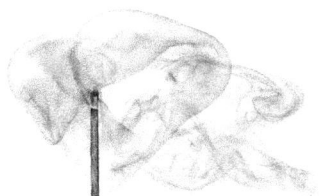

Chapter 41
BECOME THE MENTOR

In chapter 12, in the section about "Improving Yourself," we spoke about the important positive effects of having a mentor or allowing yourself to be mentored by various means. The subject is worth bringing up again here in this section on "Inspiring the Next Generation" of firefighters and leaders.

Speaking with those you lead about the importance of having a mentor can make a great difference in their lives. The greatest however, would be to work with them as a mentor yourself, and teach them to reach out and mentor others as well. Then you are not only helping that person but also the entire department. You've probably have heard the principle, "Give a man a fish and you feed him for a day but teach him to fish and you feed him for life." I want to add one more thing to that principle. "Give a man a fish and you feed him for a day, teach him to fish and you feed him for life, **teach him to teach others how to fish and you feed the entire island.**"

The importance of this kind of teaching is especially true if you are the first line of leadership for the person. In this position, you will be interacting with them daily and throughout the year, there can be no greater influence. You can make an enormous difference in their lives, the leadership of the department and the interactions with those we serve. In fact, you can do nothing better for the department and those you serve than to mentor the next generation of leaders as your replacements.

It bears repeating what former Navy SEAL Jocko Willink said, "The goal of all leaders should be to work themselves out of a job. This means leaders must be heavily engaged in training and mentoring their junior leaders to prepare them to step up and assume greater responsibilities. When mentored and coached properly, the junior leader can eventually replace the senior leader, allowing the senior leader to move on to the next level of leadership."[1]

How do you mentor others:

- **Genuinely care for them** – when you truly care about someone you are mentoring you will want the best for them. This will help you to develop empathy. Empathy is the action of understanding, being aware of, being sensitive to, and vicariously experiencing the feelings, thoughts, and experience of an individual without having the feelings, thoughts, and experience fully communicated in an objectively explicit manner. Imagine an individual is like a locked box and you need to open it to succeed. Empathy is the key to that lock.

- **Always set the right example** – people learn much more from what you do than what you say—actions speak louder than words. They will learn from your courage to always do the right thing; they will see your empathy and care for those you lead and those you care for (our patients). They will learn to put into action the words that you are teaching them.

- **Everyone is different** – not only is the individual you are mentoring different than you, but each individual you mentor throughout your career will be different than the last. Do not make any assumptions about them and do not try to make one approach fit all, this is where it takes time to get to know them.

- **Personally get to know them** – set up a meeting with them as soon as you can. Take a real interest in who they are: what are their likes, dislikes, dreams, goals, where do they want to be in five years, 10, 15, and the end of their career. What are their fears, their strengths, their weaknesses? When you really get to know them, it will help you set the path in order to help them. You will be able to strengthen their weaknesses and build on their strengthens. You will help them gain more confidence as they achieve their goals.

- **Communication** – let them know and see that the lines of communication between the two of you are always open, that you want to hear from them and help them whenever the need arises. Earn their trust and build their confidence that they can be open with you. Communication may be the most important part of the mentor-mentee relationship. It is very hard to succeed if you don't know what they are thinking or if they cannot come to you for advice.

- **Look for ways to help them** – look for situations or even create scenarios to help them develop their skills. Discuss emergency calls you handled together: what did they see, what did they learn, how would they have handled it. Do the same for non-emergency situations in the station, with the crew, or with those we meet in the public. Draw out their thoughts and help them to see and understand the right course. Look for classes or seminars for them to take. Perhaps you can go to a class or seminar with them. Introduce them to other firefighters or officers who might help them progress: remember, an individual's associations have great influence upon them.

- **Let them make decisions** – as they begin to progress, give them responsibilities. Allow them to start making decisions. Perhaps start off with allowing them to run a training exercise or something else in the station. Graduate to emergency calls. We run many emergency calls that are not really emergencies. Allow them to begin to run the scenes of calls like these; as they handle them well, allow them to progress to more important calls. You are always on the scene and can always assume immediate control if needed. This will truly help the person gain confidence as a leader. I was a very new firefighter who still needed a lot of training when, one day, I hit the three-year mark and was told that from then on, I would be stepping up as an officer when needed. The absolute sole leadership experience I'd had at that point was when I was thrown into that position early on the call about the dummy missile on the beach that I wrote about in the opening of this book. Needless to say, I had no confidence in my skills.

- **Celebrate the successes** – as the person you are mentoring continues to progress, celebrate their successes. Commend them for the good actions they take, the mindset they are developing, and when they complete a job well. People thrive when they receive appropriate praise. You are building their confidence, reinforcing good behavior, and keeping them focused and motivated.

- **Be there for the long term** – as an individual gains more insight and understanding, they will gain more confidence. As this happens, they will rely upon you less. This is a great thing: It means you have done your job well, but still, be there for them when they need you. Because you are years ahead of them in grade and on the department, they will recognize your advanced experience in many areas and look to you for advice. Being willing to share advice also reinforces the notion that you do really care for them and that is priceless.

When you choose to mentor someone, it should be a priority. Try to ensure that it's part of each day that you work together. The amount of time involved will, of course, be more intense at the beginning of your mentoring than it will when the mentee has gained experience.

The fire service work schedule is very unique, for us in Florida, we average two 24 hour shifts a week. So that means you are with your mentee, at best, for two days a week. Make them count. Use every day. When we work to mentor a firefighter, it is like planting a seed and each day afterward is like watering and caring for that seed. Seeds give us an excellent way to understand the relationship of mentoring. They are so small, sometimes almost insignificant, but they pack such incredible power. Not every seed planted will grow but when one finally does, it's transformative and it too has the ability to produce many seeds.

If you spend a good amount of your career mentoring others, imagine the effect on the department you will have. We spoke earlier about the fire service's traditions, some of which started hundreds of years ago and are still working. When you mentor others, you may well be establishing new traditions that will benefit the department and stay in place perhaps for another hundred years or more. Long after you are retired, your mentoring will still be reaping the benefits of what you have sown for those you helped, for the department, and for the people the department serves. What an incredible gift to give back to the fire service.

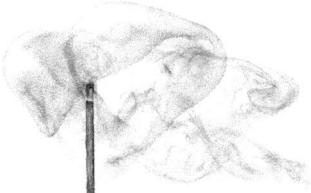

Chapter 42
THE MODERN EMPLOYEE / THE YOUNGER GENERATION

I was hired in the department in the early 1990s. As I got to know the other employees, especially the senior personnel, I continually heard the same comments about my generation and I guess, me by implication. We were the "me generation," "lazy," "entitled," "selfish," "demanding," and on and on. That same old script hasn't changed.

One day I had agreed to work a shift for the firefighter who normally relieves me. His officer, who often spewed that same old rhetoric, caught me in the hallway and told me he was going to "teach me how to work" next shift. I chuckled to myself and said "okay." I was literally a bull when I was younger, so I was not concerned. That next shift started with us going out and twisting fire hydrants.

As a department we used to test every fire hydrant in our city each June. It was a process that required opening and closing the hydrant three separate times for three different test results. Despite the intense Florida heat in June, you still had to wear your bunker pants and boots. Of course, this officer had me do all the work while he stood there with a clipboard and wrote down the results. Halfway through our list of hydrants, he didn't even bother getting out of the engine anymore. When we'd pull up to a hydrant, I would get off the engine and do all the testing myself. I'd come back when I was finished, give him the results, and we would drive to the next hydrant just to repeat.

The driver, another veteran, stayed in the engine the whole time just driving from hydrant to hydrant. He never once got out to help. It took me about three and a half hours to do them all. Crews normally break that job up over a few days, but this officer clearly was going to try to run me into the ground. It was the second half of a 48-hour shift for me, and by the time we finished I was filthy and soaked with sweat. We went back to the station, got cleaned up, and I came down for lunch. I ate by myself. I could not find the officer or the driver.

After I was done, I went to look for the other two. The driver was in the TV room sleeping in a chair and the lieutenant was in his underwear in bed watching TV. It was only 1330 hours. Since we had to work until 1700 normally, I asked him what was next. He told me to just go downstairs and study. I did as he stated. At 1600 hours, the earliest we were allowed to work out, I went to the officer to see if he wanted to join. (This was not something I should have done; but I was trying to rub his face in his "teaching me what work is" attitude. In addition, my feelings were hurt, he was not trying to help me, a new firefighter, he was only trying to cause me trouble. He was a lieutenant for the fire department, it was a position that deserved reverence and respect to me, but his actions and demeanor tarnished that position in my eyes)

Still lying around out of uniform watching TV, he declined. I told him I'd be in the gym for the next two hours if he changed his mind or needed me. As night came, these two guys were in bed by 2000 hours. I thought to myself, "And I am the lazy one? Geez, these guys are pathetic." I actually lost a lot of respect for the officer and that driver that day, and previously I liked them. It wasn't because he did practically nothing and then rested all day, but because he insulted me and then tried to bury me physically—for what reason? I thought, is that the way you lead, to try to ruin someone's reputation?

Do you think he said anything positive to me or others after seeing my performance? No, he didn't. But what would have happened if I wasn't able to keep up? Certainly, he would have broadcast that to everyone and I would've had to overcome that on my own as a very new firefighter in the department. He didn't care about me as a firefighter, as a brother, as a person, or even as an employee.

How can you respect someone in a leadership position like that? Is that what the town residents hired him to do? Would you say he served his employer well? If you owned a business, would you want a manager who tried to destroy other employees, or would you want a manager who helped them excel and taught them to provide great customer service? I think the answer is obvious. He espoused the same old tired script, "The new generation sucks, they are not as good and amazing as I was." As my career progressed and I got older, I began to hear the same things stated about the next generation as they were hired. The "Millennials" have had it the worst in my opinion[1], and "Gen Z's" often don't even get the respect of being recognized as their own generation, they're just lumped in with the Millennials.

Why do we keep showing such disrespect for people? Why do we expect newly hired young people to perform like 15-year veterans instantly? And why are these viewpoints so pervasive? Are we taught to have negative opinions about others?

A paragraph that I came across in a book on leadership I was reading can help make my point:

"We need leadership badly in our organizations because of the type of people in these organizations. People today are far more difficult and demanding, far more impatient and selfish then they have ever been before. It is no longer enough just to give a person a job and to tell them what to do. People want to participate. They want to discuss their jobs. They want regular feedback on their performance. They want to know, "What is in it for me?" Today, more and more, when people go out to look for a job, especially talented people, and members of Generation Y, instead of them approaching their job search politely and obsequiously, their attitude is, "Why should I work for you?"[2]

When I read this paragraph, I thought for sure the authors were kidding or being sarcastic. I have great respect for the authors, and I understand that this is their opinion, but it's worth a closer look. In the field of leadership there are a lot of opinions, and of course, the following is just *my opinion*. So, why am I exploring this excerpt?

Well, it helps to illustrate not only what many people are apparently thinking, but even what some in the field are teaching and thus why others are walking away with this attitude.

"We need leadership badly in our organizations because of the type of people in these organizations." That is the only sentence in the paragraph I agree with, but I don't agree with the author's premise. Their view seems to be that the problem is caused by those being led—the new people—and if a leader fails, it's the new generation's fault. In my view, the responsibility is on those doing the leading. Their statement gives a leader an immediate excuse when they fail, it's this new generation. I'm not suggesting that subordinates are never wrong, but I am saying that if you are promoted to lead, it's up to you to help your subordinates do right. It's your job to mold them into the next leaders, and not blame them for problems simply because they are young.

The authors' familiar claims that *"people today are far more "difficult,"* *"demanding," "impatient," and "selfish"* have been made about every generation since World War II and probably even before that. The same things were stated about your generation when you started and yet somehow you made it to a leadership position. There are redeeming qualities in today's hirees as well. But bringing in new people with the assumption that they are going to be difficult and demanding is the wrong way to start, as it would be in any relationship.

The next time you have a problem to solve with your partner or child, try going into it with that attitude and see how it goes. I'm sure you see the point. Don't poison the well before you drink from it. Never judge anyone before you get to know them. And even then, instead of judging, try to understand them and help them work through problems. Have a positive attitude going into the relationship.

"It is no longer enough just to give a person a job and to tell them what to do. People want to participate. They want to discuss their jobs. They want regular feedback on their performance." Are wanting to participate and wanting feedback bad things? In my opinion these are exactly the ambitions you want from young staffers. For the first 14 years of my career, we had an assistant chief who dealt with personnel daily. He constantly addressed issues with this reply, "If you don't like it, hit the bridge, there are 300 people who we could replace you with." (Our town was on an island so "hit the bridge" meant "leave the job and get off the island.") This did not invoke passion for the job, nor did it make people appreciative of what they had. All it did was anger people—the solution to any issue you had was quit and be replaced.

You *want* people to be involved with their job, asking for feedback, and concerned about their performance. If you are looking for someone to build a house for you, do you want someone who just does his job? Or would you rather someone who is concerned about their performance, loves what they do, and wants feedback? You know that second person is going to do the best job they possibly can for you. For the other person, the job is just a paycheck.

What about the question of modern employees wanting to know *"what's in it for me?"* or *"why should I work for you?"* Again, I do not see the problem here. Are they saying having a job is such a privilege that the employee should just shut up and be grateful? I believe it *is* a privilege and we *should* be grateful, but something like owning a home is a privilege too. So, when you go to buy a house does that mean you should just accept what the seller is asking no matter how high the price or how many problems the house has? Or do you negotiate? Do you ask why you should you buy this house? Is it the best for your family? When you get a loan for a house, do you accept the terms no matter what? Or do you shop around for what is best for you? What's in it for you? When people today come into a job with higher expectations, hoping to get something out of it, I don't believe it's a bad thing.

The very next line actually read: "*One of the major reasons that people go to work for any organization is because of the leadership.*" EXACTLY! Who would want to work for an organization where the leadership assumes you are *difficult, demanding, impatient,* and *selfish?* Or one that adopts the attitude that they are just placating an entitled child? No one, and neither would you.

Good leadership means that you take people of every age group, every gender, every race and background, and you mix and meld them into a successful team. Everyone is equal and appreciated for the gifts they bring. You help them achieve success individually and as a team. You help them appreciate that they matter, that they make a difference to the team, and more importantly—to those we serve. You let them know that you want them there and you want them to be very successful. It doesn't mean that you let them get away with whatever they want to do. The fire service doesn't work that way.

So then, how do we lead this new and younger generation or anyone for that matter?

1 - Understand

Don't label anyone and don't judge anyone. Do people today have higher expectations of a job than in years past? Probably. Today, those entering the work force can expect more: there's been low unemployment for many years and stiff competition for employees. They are given more bean bags, dress-down days, employee cafeterias that provide excellent food, and many other benefits. They are blessed that way.

However, let me tell you what else this generation faces. More than ever, people need a college education to get a good job. Even the fire service has changed drastically in this regard. College is more expensive than ever before, so this younger generation leaves school seriously in debt. We didn't. More than ever, companies do not have retirement pensions, they have 401(K)s which are not nearly as good as a pension.

The country that is supposed to care for them when they get old has more debt now than at any time in history. Younger generations will have to figure out a way to pay this off or suffer losses of benefits even though they will pay-in far more than you and I and anyone else did previously to receive this benefit.

When I grew up, our television had channels 2, 4, 5, 7, 9, 11 and 13(PBS). The local news came on at 6 p.m. for 30 minutes and then the world news at 6:30 p.m. for 30 minutes. That was it. Today, people are bombarded with thousands of media outlets reporting 24 hours a day, 365 days a year of all the terrible things going on in the world. Let's applaud that new generations want to make a positive difference, a positive impact.

The list can go on. So, do not judge people. Instead, try to understand their background and work with them to improve the traits they need to improve. You may actually find out that they aren't that different from you after all.

2 - Look at their goods and learn.

I appreciate the generosity, openness, and tolerance of the Millennials and younger generations. Many of them donate time and money to help others and at young ages, even before they have a house or other material things. I grew up a certain way—a rough upbringing in a rough town at a rough time. It shaped me. My kids and many others in the current generation have had a wonderful influence upon me and I am eternally grateful for that. When people see that you appreciate who they are and that you have even learned something from them, they are drawn to you and to your leadership. They will allow themselves to be more vulnerable, which will in turn allow you to help mold them for the department. If you think there is nothing you can learn from a younger person, you have already stopped growing and are regressing.

3 - Truly care and let them see you care.

This is something expounded upon in chapter 33 and touched on throughout the book. It's perhaps the most important thing you can do. If the people you lead see that you care, they will trust you, and even when they don't understand your direction or how it is given, they will trust it and follow. I saw this point beautifully illustrated in a November 2018 interview with Tom Brady. Brady was quarterback for the New England Patriots at the time, and the interviewer was Rodney Harrison, an ex-defensive football player and now sports commentator. Although I am a New York Giants fan, I truly admire Tom Brady. What he has been able to accomplish in his career is astounding. In this interview he was humble and gracious, talking about some of his accomplishments when Mr. Harrison asked him, "Have you had to adjust your leadership style in order to relate to this younger locker room and younger generation?

Brady answered, "I feel like I do that naturally," and explained that he had others to look up to when he was younger, and he tries to be like that for the younger generation.

Harrison responded, "One thing I see is you don't hesitate to yell and scream, whether you're yelling and screaming in the wind or yelling and screaming at your guys, do the younger guys take to that?"

"Yeah, they do," Brady answered, but he appeared to be somewhat uncomfortable. Then he added, "By the way, you yelled a lot too."

Harrison responded, "It was a different time then," adding, "we had a lot of veterans and the young generations are a lot different now, they're a little bit more sensitive." (Another frequently used adjective for them.)

Brady explained he wasn't yelling just to yell at his team. "Communicating in the locker room and communicating on the field are very different," he said.

Harrison interrupted, "But they trust you."

"Yeah, they trust me," Brady remarked, "and they trust I am trying to do it in all of our best interests, I'm not trying to do it to show them up. So, I have never had a problem with that."[3]

Now, I am not condoning yelling at anyone and I am not judging Tom Brady for it, that's his profession. But notice that he was able to get away with doing so for one reason—his team trusted him, and they trusted him because they knew he was doing what was best for them. He wanted what was best for them. People will work hard for you; they will do almost anything for you when they trust you and they *will* trust you when they believe you care for them.

Remember that lieutenant I told you about at the start of this chapter? The one who made me test all the hydrants while he just lolled in his seat? That experience with him left me knowing he just wanted to harm me. How could I ever trust someone like that? If you do as he did and walk into a relationship already thinking that someone is lazy, difficult, selfish, or anything negative, it will be difficult to ever win their trust. But if they think you genuinely care for them and want to help them, they will lasso the moon for you. Try it and see the results.

4 - Treat everyone fairly.

Always be fair to everyone. Don't play favorites.

There's no quicker way to lose a person's respect or trust than to treat them differently than the others. When a person feels that hard work, competency, desire, and character mean nothing, and instead they are judged based on their age, personality difference, or outlook in life, you will lose them. They will either continue in frustration and their interactions will get worse with their supervisor, or they will quit.

They will see a supervisor who treats people that way as irrational and ignorant, and they can't respect that. Justice and the equal treatment of all are beliefs that are strongly ingrained in the younger generation. If they view themselves being treated unfairly, it will be very difficult for them to accept.

Instead of treating a younger employee differently, you should spend extra time with them. Those new to the fire service will need help transitioning to a para-military organization. They need to learn procedures and rules and regulations, how to use all the equipment, how to care for the units and the stations, the treatment of patients, and so much more.

5 - Help them strive to succeed—satisfy their need to effect change.

Your responsibility as a supervisor is to those you lead. It is to integrate them successfully into the fire department, into the team, and to help them then progress, to help them reach out, and help them succeed. This leaves no room for you to talk negatively about them or do things to try and harm their career. Doing those things shows what lack of leadership skills you have and broadcasts your own inadequacies instead of theirs. If they want to feel that they make a difference, well then, they came to the right profession. We make a difference in other people's lives every day. Teach them this, teach them how, help them love the people they serve, help them love the job, help them make a difference in a world that desperately needs it.

Leading anyone can be a challenge. Each generation in our country is different than the last. The further apart in age, the bigger the differences can be. This does not mean one is better than the other, they are just a little different. But what remains the same is potential. They all have the potential to excel.

When my department used to test fire hydrants, we tested for three different pressures, the flow pressure, the static pressure, and the residual pressure. The flow pressure was the PSI (pounds per square inch) of water when it is flowing out of one of the hydrant outlets. The static pressure was the full available pressure of the hydrant. It was tested by putting a gauge on one of the outlets and opening up the hydrant. No water would flow, it would come up against the gauge which gave a reading. The final reading was the residual pressure. This was the amount of pressure that was left in the hydrant after it was already flowing. This was important to know so you could calculate how much more water you could get out of that hydrant if you were already using one valve.

Perhaps a new employee is not unlike a hydrant. They come in with loads of static pressure—skills, energy, and desire to do a great job. When they start working, they are now flowing. Hopefully you guide and train that pressure in the right directions. Never give up on them. No matter their differences or difficulties, there is always more to them. There is that residual. Work to get to that residual if need be. Try to get to a place where you understand who they are and what their desires are. Look for the good in them and help them accentuate their good. Use that good to integrate them into the department. Learn about them and get to care about them and let them see that. Treat them fairly and help them strive to succeed and you will have a successful subordinate/supervisor relationship.

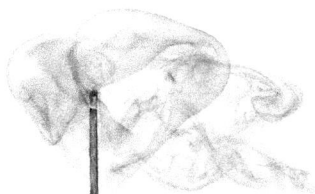

Chapter 43
MENTAL HEALTH

It was a gorgeous afternoon. Winter in South Florida can truly be described as a paradise: beautiful blue skies, bright beaming sun, and cool, comfortable temperatures. It almost seems like nothing can go wrong on those days. But, unfortunately, that was not the case this day. We received a call for a possible suicide. My stomach knotted up as we headed to the scene, hoping we could help the patient. As we arrived, the police were all over the scene and would not let us approach until they felt it was safe.

A woman had called 911 and told them that her ex-boyfriend was threatening to kill himself with a gun because she broke up with him. When the police finally cleared the scene, they called us in. I walked up and opened the car door. The smell of blood and gunpowder smacked me right in the face. There he was, head slumped over to his left, towards me, blood was splattered all over the inside of the windshield and his eyes open. Looking at his open eyes, strangely, I could see desperation, it looked like his eyes were talking to me, they were screaming for help. From a hole in the side of his head, there streamed what I can only describe as what looked like blood stalactites. We were way too late. This gorgeous day was marred by the terrible loss of a person in so much pain. To this day, I can still see every detail of him in my mind, pictures of a desperate dead man with a hole in the side of his head, and the smell of stale blood and gunpowder, a scene neatly filed in my brain and ready to be recalled at the slightest whiff of anything similar.

The image of this man is just one permanent memory of so many in my brain. I remember an individual who jumped off an eight-story building and whose body, when we went to lift it, felt like his bones and organs had been turned to liquid. Lifting and moving such a person is a jarring nightmare. It disturbs me to this day when I recall it from time to time.

There was the construction worker who stepped on a skylight that cracked. He fell nine stories to a gruesome death. Seeing a body mangled in a position it should never be in is something you just cannot forget.

I still see the frantic looks of several different individuals who cut their wrists in an attempted suicide. Blood everywhere, the person screaming and flailing about and as I help them my uniform became soaked with blood. One situation so tense, so emotionally packed, I vomited profusely after the call in the front of the person's house.

I still feel deep sadness from responding to a call for a man, who, while tenting a building for termites, touched a live wire on the electric pole and was electrocuted in the extension bucket, right in front of his brother. Images of trying to work and revive the man while hearing the guttural screams and cries of his brother and the smell of burnt skin, haunt me.

I think of the countless times I responded to an individual who died— working our hardest to revive the person while their partner, so desperate for us to work some magic, continually called out cries of, "please, please I cannot lose him." Having to finally walk over to that person after an hour and telling them their partner was dead, we'd done everything we could, still crushes me today. As they collapsed into an emotional breakdown, grabbing me and hugging me, I could never hold back the tears myself. I would always call my wife after those times, thinking of their pain and the similar pain one of us would one day experience. Sometimes I would talk, but most times I just wanted to hear her voice as I tried to push down and process the pain of what I had just experienced.

But nothing was worse than the children. Seeing an innocent child hurt, in pain, abused, or dead is unlike any other terror. Firefighters and police officers seem to have a lot of children themselves. When you see an innocent child hurt, all you can see is your own child. The images play over and over again in your head as you imagine this terror happening to one of your children, hearing their screams, seeing their pain. It can be overwhelming and debilitating for quite a while, and even then, it's always there just below the surface, ready to resurface and bring back with it all the pain, fear, and sadness when something jogs the memory.

I could write a book, not just a chapter, on all the calls that permanently scarred me and are stored in my brain. And there are many departments and firefighters who have seen and experienced so much more than me.

Unfortunately, as a firefighter, it's not easy to admit you've been mentally impacted by a call. If you get even a little emotional, you can count on word of it spreading throughout the department and some people mocking you behind your back. After a baby drowned in a backyard pool, the lead officer on that scene had trouble processing the event. There was nothing in place to help him. After some time without getting any better, he was advised to see a professional and took a few shifts off. I remember him being viewed as soft and fragile within the department. Everyone thought he was going to quit. He never seemed to be the same after that incident. He was more serious, quick to get upset, angry, and had a sad countenance most of the time.

Years later, we would come to understand that he was experiencing PTSD (Post Traumatic Stress Disorder). In time, PTSD finally became a topic for serious conversation and priority consideration in the department. A procedure was typed up and people were told that counseling is available. Then, just as quickly as the topic arrived, it dropped and was never brought up again. The department never challenged the negative attitudes toward seeking help. To my knowledge, no one used the services willingly. Why would they? It would destroy their reputation and possibly their opportunities for promotion. This is one of those places where tradition in the fire service does not serve us well.

Firefighters have a very tough physical job and often we get hurt. We have no problem documenting physical injuries and taking time off for such injuries, but when it comes to mental difficulties from our job, or even other events in our life, we are not very forthcoming because we have accepted the stigma that goes with it. That stigma at times can even be used and wielded like a weapon, ready to destroy a person.

WE HAVE TO GET RID OF THAT STIGMA

Not recognizing mental health issues is dangerous to the individual firefighter and to the fire service as a whole. That is why as a leader you must address this situation. Your department may already have some procedures, and that is fantastic. Is the topic of mental health brought up at least yearly for conversation? Is real science used to help people understand that not only is it okay if you are emotionally affected by our job and the things we see and go through, but that it is imperative that these things are addressed? That the long-standing effects could damage a person in many ways and resurface throughout his or her life, especially when they get older? That their personal relationships with their partners, children, family, and friends can be negatively impacted and even damaged beyond repair?

These incidents can cause flashbacks that are so real that you see, feel, and even smell the incident all over again. They can lead to night terrors, anger, bad dreams, avoiding people and places, sleeping difficulties, emotional outbursts, sadness, being on edge, memory problems, guilt, mood swings, and a host of other issues. You can be going along fine one moment and then suddenly are hit with strong emotional feelings that you have difficulty controlling.

You are the leader. As we've gone over many times now, you need to have special care, even love for those you lead. That means you need to be aware of what's at stake and set a good example. That means seeking and accepting help for your own mental health as well.

By not accepting help and instead trying to act like Superman or Superwoman, you do two things. First, you give the impression to those you lead that counseling is for weak people. It helps keep alive the stigma against getting help. That is the exactly the opposite of the impression you want to give: it defeats the purpose, and you hurt those you lead instead of helping them. Second, you are personally weakened by the experience.

If you were experiencing chest pain every time you exerted yourself, wouldn't you go to the doctor to get it checked out? If you do not, certainly you can have a heart attack which can either kill you or weaken you significantly. Your entire life will be altered. However, if you were examined by a medical professional and they addressed that problem, you would be back to normal, perhaps with a few lifestyle changes like more exercise and eating better. The same is true with traumatic events. Not addressing them will lead to long-term damage, whereas addressing them with a professional will help you heal.

You want those you lead not only to succeed but to excel. Help them to see the positive effects of addressing the situation after they have experienced a traumatic event as well as the need to look out for others and helping them too, to see the utility and necessity of getting help. You want to ensure that those you lead and others in the department do not think of help for PTSD as being soft or weak.

Get a clear education on the signs and symptoms of PTSD. The earlier the problems are recognized and addressed, the better the results. Of course, you want to act within the guidelines of your department. If help or treatment is not presently available or highly recommended, you might start by bringing the needed information to your superiors.

Interestingly, throughout my career, I had to constantly study our medical standard operating procedures. In this large book were signs and symptoms to look for in our patients, so we could decipher what issues they were facing and, when we did, know what medications or procedures to use to help them. Many procedures would change almost yearly as medical science evolved better ways to treat people. We would be tested on our knowledge and if we failed, we could lose our paramedic license. Yet all the while, there was nothing out there for leaders to use to assess signs and symptoms of mental issues that may have begun to appear in our own team members. We need to be as well equipped to help those we lead and serve side by side with as we are for our patients.

If your department already addresses this subject, I applaud your leaders. It is still up to you to understand PTSD and become very familiar with its circumstances and signs. If your department doesn't have a program for that, it's still up to you, staying within your department's guidelines, or, even better, helping your department to see the need and fulfill it. In this way you are not only inspiring the next generation and helping them to succeed but you are protecting them as well.

Information about the diagnosis and care for PTSD is rife with stigmas, myths, misrepresentations, and falsehoods. Let's get a better understanding of what it is and how to spot it. Some basic facts follow:

WHAT IS PTSD?

Post-Traumatic Stress Disorder is a trauma and stress-related disorder that may develop after exposure to an event or ordeal in which death or severe physical harm occurred or was threatened.[1] It is a disorder that can be debilitating to those who suffer from it and destroy not only the life of people who suffer from it but also those who love them.

WHY IS IT AN IMPORTANT TOPIC FOR US TO DISCUSS?

Due to the nature of our job—the traumatic events that we encounter many times throughout our career—firefighters have much higher rates of PTSD than other professions. To make matters worse for firefighters, many fire departments around the country handle all medical emergencies as well. Studies have found that firefighters generally report that medical emergencies and motor vehicle accidents were the most upsetting types of calls that they received.[2]

This works as a compounding effect and increases the probability of suffering from PTSD significantly. The International Association of Firefighters (IAFF) in a 2016 report estimated that 20% of firefighters and paramedics exhibit symptoms of PTSD as compared to 3.5% for the general population.[3] Other studies have found that between 7% and 37% of firefighters meet the criteria for diagnosis of PTSD.[4] Besides the traumatic events we handle here are some additional risk factors:

- Starting as a firefighter at a young at age
- Being unmarried
- Holding a supervisory rank
- Holding negative beliefs about oneself (that you are inadequate or weak.) (The perceived stigma of mental health issues will make this worse.)

HOW DO WE RECOGNIZE PTSD?

Symptoms can be exhibited both physically and/or mentally. Headaches, chills, sweats, panic attacks, and heart palpitations are some physical effects. In addition, the following is a list of symptoms found throughout many medical sites.

- **Re-experience** – nightmares, flashbacks, manic episodes, and persistent sad memories, short term memory loss
- **Avoidance** – avoidance of people, places, and items that remind the person of strong feelings, good or bad
- **Constant negativity** – person may develop a negative view of the world and everything around them, a sense of injustice and betrayal, a desire for revenge
- **Hyper-reactivity and arousal** – easily startled by loud sounds, difficulty sleeping, difficulty resting, sleep deprivation
- **Arousal** – increased anger, aggression, or irritability, lack of interest in sex, sexual dysfunction
- **Mood changes** – anxiety, depression, mood disorders, feelings of

detachment and guilt, lack of interest in previously enjoyed activities, distorted beliefs about oneself, others and the world. [5]

- **Numbness** – lack of initiative or ambition, malaise, lack of interest in past hobbies, becoming overly engrossed in TV or the computer.

To make it even worse, PTSD can act as an exacerbator and put you at much greater risk of developing a number of other mental health disorders. If you experience PTSD, you are...

- Five times more likely to develop an anxiety disorder
- Six times more likely to develop depression
- More likely to develop an eating disorder
- More likely to develop substance abuse disorders
- Six times more likely to commit suicide[6]

These numbers are unacceptable, we should not let our brothers and sisters be affected this way by something we can definitively address and fix.

PTSD is extremely dangerous and, often, people who struggle with it do so in isolation. But just because you have experienced a traumatic event and will again in the future, does not mean you will definitely develop PTSD.

CAN YOU DEVELOP PTSD FROM A SINGLE EVENT?

It is certainly possible, especially if it is not handled correctly. However, although it is usually not necessarily the case, a single event can have lasting damage. More likely, events can have a cumulative effect, and if not addressed initially, they can build upon themselves and eventually lead to PTSD. We can illustrate it this way:

First, each person deals with things differently. Second, every event affects you differently, in fact the same event will affect different individuals on the scene differently. In addition, the degree to which someone is affected will differ based on their past life experiences. Finally, a lot has to do with how much prior stress the individual is carrying.

If you were lifting weights and you decided to grab a 15-pound weight and curl it, it would perhaps be super light for you. Perhaps you could curl it over and over and over again without any strain. But if you keep doing it without taking a break, it will start to get harder and harder and eventually you will not be able to lift it at all. If you do not stop to rest and recover, the weight, just 15-pounds, will become too heavy for you to handle. Likewise, during the course of your career, you will face traumatic events one after another. In the beginning, it may feel easy to handle them, but you need to take the right steps to rest and recover from those events. Otherwise, they will build up and you will be unable to continue.

If you do experience PTSD, are there treatments? The answer is *yes*. First there is preventative care.

- *Social support available either at home or work* – having access to be able to talk about incidents and how the person is feeling is an excellent way to begin to heal.
- *Expressive writing* – In the book, *Opening Up by Writing it Down*, James W. Pennebaker, PhD, and Joshua M. Smyth, PhD, elaborate on many studies that have shown that using expressive writing techniques have helped people manage PTSD symptoms, change emotional states, bring changes in health styles and health behaviors, improve mood, experience greater post traumatic growth, reduce psychological stress reactivity in response to traumatic memories, help individuals understand the event and themselves better, have fewer nightmares, better sleep, lower blood pressure, and more.[7] A warning though, different people react differently to expressive writing just as they react differently to the event itself. Sometimes people initially feel somewhat sad or depressed before getting positive effects.[8] It may be best to ask a professional about

this preventative therapy before using.

- **Telling your trauma story** – similar to writing it down, it can be incredibly healing to talk about the trauma with a trusted person. Seth J. Gillihan, PhD, expounds on the effectiveness of this step. Feelings of shame may subside, unhelpful beliefs about the event can be corrected, the memory becomes less triggering, you find a sense of mastery, the memory becomes more organized and you begin to make sense of the trauma.[9] If you wish to act as this trusted confidant, or if the person sees you as that, you must keep their words private, you cannot destroy their dignity or trust in you. Doing this may close them off to trusting anyone and cause them to keep all their feelings to themselves. This only exacerbates the problem. However, this trusted position should never hinder you from recommending professional help.

- **Effective coping strategies**—preferably outlined in city or department policies, procedures, and regulations.

- **A culture of openness and honesty** – A firefighter must believe they can get help without penalty, they must feel and believe in the protection of their department's program, they are going through a very vulnerable time: the feeling of protection is key. Confidentiality is a must. In his book, *The Body Keeps the Score*, Dr. Bessel Van Der Kolk writes, "Being able to feel safe with other people is probably the single most important aspect of mental health."[10] If not handled professionally or correctly, individuals have turned to unhealthy methods of self-medicating, including drugs and alcohol. Some even attempt suicide. Here is where your leadership can really shine, or it can fail miserably. As we discussed earlier in the book, caring for those you lead and letting them see that will make an individual more likely to come to you and express the way they are feeling. This in itself can save a firefighter's life.

- All leaders should be trained to recognize symptoms, *though the diagnosis and treatment of PTSD is to be done only by professionals.*

Officers who are great leaders carefully watch over and care for those they lead. They are trained to look for signs and symptoms in those they lead. They are trained to be able to be the first line of social support and trained to fully know and understand the policies of their department.

- Some women, members of diverse ethnic groups, and members of the LGBTQ community have experienced exclusionary behaviors that made it difficult for them to feel like a member of the fire service family. Sometimes they receive such flack that it puts them on the defensive. Many of the women I have worked with have described it this way. They have stated that they feel they have to constantly prove that they can do the job. If any firefighter feels this way, they are even more likely to not come forward when they are having a difficulty with mental health. Thus, as a leader you need to be even more vigilant in watching for this. I have learned much from the women I worked with in the service. It made me a better person and a better leader. In fact, the best officer I ever worked for was a woman. I am grateful I had the opportunity to work for her.

WHAT ARE SOME OF THE TREATMENTS AVAILABLE?

- Cognitive behavior therapy
- Exposure therapy
- Cognitive restructuring
- Cognitive processing therapy
- Eye movement desensitization and reprocessing
- Stress inoculation training

In addition, there are preventative measures that can be taken before you encounter a traumatic event, that will make you better at handling the pressures and strain that will come. These same preventative measures can also be used to help you successfully deal with a traumatic event once it has happened. A survey was conducted with 225 individuals who had escaped the twin towers. They were asked what helped most in overcoming the effects of the experience. Acupuncture, massage, yoga and EMDR, in that order, were mentioned by those survivors as what helped them the best.[11]

As I am in the middle of the second edit of this book, an acquaintance of mine, a fellow firefighter brother from a different city, a guy I talked with and have gotten to know well as we both worked out at a local gym for several years, just took his life. He shot himself while he was on duty at the station. He was young, had a full life ahead of him, he was in incredible shape and seemed to be a very happy person. You can never know what a person is hiding behind their smile.

Did he give off warning signs? Did the pandemic or other personal factors in his life contribute to this action? I don't know. I had not seen him in a year, since the pandemic started. I do know he was a good man. He helped and served many people. He was very well liked. It is a devastating loss and it is a loss that happens too often in the fire service. We have to do a better job at taking this topic very seriously and educating each other about the dangers of PTSD. We have to do a better job of killing the stigma, that it is weak to seek mental help. We have to do a better job at identifying individuals experiencing this pain and helping them through it. We have to do a better job at serving and protecting our brothers and sisters in the service. It is my hope that this book and additional work will contribute to the health of the fire service and all those that serve in it.

COMMENCEMENT

It is so exciting to be here at this point in the book. But should this chapter be called "conclusion" instead of "commencement?" We have indeed come to the end of this book, but it is not the conclusion, it is the beginning of your journey. It is time for you to commence your journey, even if you are already in a position of leadership. Your journey should take a lifetime, never stopping. You should never stop reading, never stop learning, never stop growing, never stop improving. Make this just the beginning of change, the beginning of improvement, and keep pushing forward. Make it a journey you will be proud of.

For a moment, I want you to go on an imaginary trip with me. I want you to think of something you wish you had or wish you did not have. Let me give you some examples. Perhaps you may wish to be in much better shape, maybe you wish you could run 10 miles or even a marathon or have a lot more muscle mass. On the flip side, perhaps, you may wish you did not drink soda, or eat so much bad food, or drink so much alcohol. Maybe you wish you spent more time with your family, or you were still taking college classes, or you read more.

Whatever it is, now imagine for a moment that you started working on that a year ago. Where do you think you would be right now? Now imagine you started five years ago. Think for a moment where you would be right now. Meditate on it for a few moments, what your life would look like. Put it in your mind where you want to be, what you want to be or what you want to accomplish. Now do it. Do not wait. Do not wait another day. Start game planning. Start today.

Now imagine for a moment where you will be in a year if you stick to your plan and work hard. You might have heard the aphorism, "The secret to getting ahead is getting started." If you do start today rather than a year from now, you will be able to look back and see significant progress and change and in five years your life may be completely different. Tim Fargo, author of *Alphabet Success*, popularized the phrase, "Who you are tomorrow starts with what you decide today."[1] That is a fact, for the good and for the bad.

This material and your desire should help you get started, move forward, and keep moving toward all your desires. Make goals for yourself: small goals, goals that are attainable. As you achieve your goals, make new ones and larger ones. Before you know it, you will be able to look back and see good progress. When you fail, and you will, don't give up. Get right back up and keep moving forward. Do not ever make excuses for any failure or setback. Mr. Fargo also says, "Excuses are the rocks where our dreams are crushed."[2] Excuses will only make you give less effort and try less often and eventually give up, not achieving that goal you wanted.

As you go forward, remember that life is amazing, and you have the best job in the world. In addition, you have an opportunity to make an impact, a life altering impact, for the benefit of others. Never forget that the foundation of the fire service is *Service to Others*. We do this not only for those we respond to but also to those we lead and those we set an example for.

There's a quote I like, often attributed to George Bernard Shaw, that says, "We are made wise not by the recollection of our past, but by the responsibility for our future." We have a responsibility to those we serve. We have a responsibility to our country. We have a responsibility to care for the future fire department. We have a responsibility to humanity. If we serve others and we set the example and help others to learn to serve as well, we are well on our way to a better planet.

Use this book as a continual reference. Come back and re-examine areas as you need them. Reread the book and you will get more out of it. Always care for your customer like they are your own mom. And above all things always give Love!!

When accepting his Nobel Peace Prize in 1964, Dr. Martin Luther King, Jr., said "I believe that unarmed truth and unconditional love will have the final word in reality."[3]

Give your love to as many people as you can. Love is the final word.

Love!

x

Appendix A
RECOMMENDED READING

The Autobiography of Martin Luther King, Jr., edited by Clayborne Carson

Awaken the Giant Within: Take Immediate Control of Your Mental, Emotional, Physical And Financial Destiny, by Anthony Robbins

Biggest Brother: The Life of Major Dick Winters by Larry Alexander

Can't Hurt Me: Master Your Mind and Defy The Odds, by David Goggins

Daring Greatly: How the Courage to Be Vulnerable Transforms the Way We Live, Love, Parent, and Lead, by Brené Brown

Extreme Ownership: How U.S. Navy Seals Lead and Win, by Jocko Willink and Leif Babin

The Happiness Hypothesis: Putting Ancient Wisdom and Philosophy To The Test Of Modern Science, by Jonathan Haidt

I Never Had it Made, an autobiography, by Jackie Robinson

In Order to Live: A North Korean Girl's Journey to Freedom, by Yeonmi Park with Maryanne Vollers

Life Without Limits, by Nick Vujicic

Man's Search for Meaning, by Viktor E. Frankl

Maximum Achievement: Strategies And Skills That Will Unlock Your Hidden Powers To Succeed, by Brian Tracy

Modern Man in Search of a Soul, by C. G. Jung

My Experiments with Truth, by M. K. Gandhi

Open Up by Writing it Down: How Expressive Writing Improves Health And Eases Emotional Pain, by James W. Pennebaker, PhD and Joshua M. Smyth, PhD

Principles, by Ray Dalio

The Body Keeps the Score: Brain, Mind, And Body In The Healing Of Trauma, by Bessel Van Der Kolk, M.D.

The Complex PTSD Workbook: A Mind-Body Approach To Regaining Emotional Control And Becoming Whole, Arielle Schwartz, PhD

The Ride of a Lifetime: Lessons Learned from 15 years as CEO of the Walt Disney Company, by Robert Iger

Start With Why: How Great Leaders Inspire Everyone To Take Action, by Simon Sinek

12 Rules for Life: An Antidote to Chaos, by Jordan Peterson

Win the Heart: How to Create A Culture Of Full Engagement, by Mark Miller

Appendix B
SUCCESS CHART

Here is a copy of a chart that I use to help me either work on a deficit I feel I have or something I want to add to my life. I hope it can help you. Download a free copy at www.fdleadership.com

	M	T	W	T	F	S	S
Week 1							
Week 2							
Week 3							
Week 4							
Week 5							
Week 6							
Week 7							
Week 8							
Week 9							
Week 10							

Appendix C
LOVE

We've spoken about loving your job and caring about those you lead and those you serve. But I wanted to include a thought just on love, to implore you to give love and reason with you on the importance of our showing love to everyone. It will benefit us and all of humanity, there is no downside to it.

I have read and studied about leaders and leadership principles and qualities for over three decades. I have learned so much and I am eternally grateful for what it has done for me in my role as husband, father, son, and friend as well as in my career. Many times, leadership and life itself is presented as a cold and surgical list of rules of do's and don'ts or axioms learned from the example of a successful leader in a certain field. But I believe that one quality is interwoven throughout all of leadership and life itself and holds all the other stitches in the cloth of life together. This quality is love.

When we lead, we lead people and people need something extra. Let me illustrate. I have been investing in the stock market since I was a kid. When it comes to investing, the greatest minds will tell you that one of the most successful things you can do is to take any emotions out of it. The better you are at doing that, and sticking to a list of disciplined rules, the better investor you will be. Perhaps you have heard the saying "buy stocks when there is blood in the streets," which means that one of the best times to buy stocks is when everyone else is panicking and selling everything, when the market is dropping precipitously. Historically, if you've bought at those times, taking emotions out of it, history has shown that you would have been very successful.

The idea of following emotionless principles works not only in investing, but in many areas, including the fire service. For example, when we see a child in a precarious situation, crying, bleeding, hurt, and suffering, we cannot afford to let emotions cloud our emergency treatment protocols. When we have to enter a building to save lives and everyone else is running away, we cannot let emotions or feelings take over. We have a job to do. But leadership is so much more than just the emergency situations.

In fact, it's in everything leading up to the emergency situations where leadership is built. The entire premise of this book is to show you that a leader displays these qualities *at all times* and is working to improve *at all times*, in the little jobs, the training, the cleaning, the minutia—and during all those times you are leading people with emotions and feelings.

Therefore, while I believe in a disciplined approach, and a list of rules or goals works well for us when we work on ourselves, when we lead others we must take into account their feelings and allow our own feelings to show.

Love is the greatest feeling, the greatest quality you can give to others. It's the quality that makes life amazing and exhilarating, a divine gift that brings you happiness and success.

Can you picture a world filled with love? We are the pieces to put it together, we are the ones to interweave it throughout society. German Poet Johann Wolfgang von Goethe wrote, "Love does not dominate; it cultivates."[1] Cultivate love wherever you are and with everyone around you. You will have the kind of effect that you cannot even imagine. There's a beautiful quote, often attributed to American poet and civil rights activist

Maya Angelou, that says "I've learned that people will forget what you said, people will forget what you did, but people will never forget how you made them feel." People will never ever forget the love you expressed for them.

What is love, how do we express it? The Bible defines it this way, "Love is patient, love is kind. It does not envy, it does not boast, it is not proud. It does not dishonor others, it is not self-seeking, it is not easily angered, it keeps no record of wrongs. Love does not delight in evil but rejoices with the truth. It always protects, always trusts, always hopes, always perseveres. Love never fails."[2] Print that out and carry that with you each day. Can you imagine if every person in your life or people you interact with treated you that way? Set the example. Do it first to others and let them learn from you. Let them feel that joy and happiness and success from your love and they will want to repeat that love back to you and others.

So, I encourage you to make love the primary focus of your life. Love and care for others. Love yourself. Love your job. Love those you lead. Love those you serve. Victor Frankl, after seeing the worst of humanity and the best wrote in *Man's Search For Meaning*, "A thought transfixed me: for the first time in my life I saw the truth as it is set into song by so many poets, proclaimed as the final wisdom by so many thinkers. The truth—that love is the ultimate and the highest goal to which man can aspire. Then I grasped the meaning of the greatest secret that human poetry and human thought and belief have to impart: The salvation of man is through love and in love."[3]

"Love is all you need."
— The Beatles

Appendix D
PTSD RESOURCES

American Psychological Association – Resource guide for traumatized people and their loved ones.
http://www.apa.org/topics/trauma/

Trauma Center at JRI. (associated with Bessel Van Der Kolk, M.D.)
http://www.traumacenter.org

Gift from within. PTSD Resources for Survivors and Caregivers.
www.giftfromwithin.org

HelpPRO Therapist finder. Comprehensive listings of local therapists specializing in trauma and other concerns.
http://www.helppro.com

PILOTS database at Dartmouth is a searchable database of the world's literature on post traumatic stress disorder, produced by the National center for PTSD.
http://search.proquest.com/pilots/?accountid=28179

GOVERNMENT RESOURCES

National Center for PTSD includes links to the PTSD Research Quarterly and National Center division, including behavioral science division, and clinical neuroscience division, and women's health sciences division.
http://www.ptsd.va.gov

National Institute of Medical Health
Http://www.nimh.nih.gov/health/topics/post-traumatic-stress-disorder-ptsd/index.shtml

PROFESSIONAL ORGANIZATIONS FOCUSED ON GENERAL TRAUMA RESEARCH

International Society for Traumatic Stress Studies
www.istss.com

European Society for Traumatic Stress Studies
www.estss.org

International Society for the Study of Trauma and Dissociation
http://www.isst-d.org

PROFESSIONAL ORGANIZATIONS DEALING WITH PARTICULAR TREATMENTS

The EMDR International Association (EMDRIA)
http://www.emdria.org

Sensorimotor Institute (founded by Pat Ogden)
http://www.sensorimotorpsychotherapy.org/home/index.html

Somatic Experience (founded by Peter Levin)
http://www.traumahealing.com/somatic-experiencing/index.html

Internal Family Systems Therapy
http://www.selfleadership.org

Pesso Boyden System Psychomotor Therapy
www.PBSP.com

THEATER PROGRAMS

Urban Improv
http://www.urbanimprov.org

The Possibility Project
http://the-possibility-project.org

Shakespeare in Court
http://www.shakespeare.org/education/for-youth/shakespeare-courts/

YOGA AND MINDFULNESS

http://givebackyoga.org
http://www.kripalu.org
http://www.mindandlife.org

ENDNOTES

Chapter 1

1 Laozi, *Tao Te Ching* (New York: Penguin, 2000).

Chapter 2

1 Jay Luvas, *Napoleon on the Art of War* (New York: Simon & Schuster, 1999).

2 Marcus Aurelius, Mark Aurel, trans., *The Thoughts of the Emperor Marcus Aurelius Antoninus* (Meditations) (London: George Bell & Sons, 1891).

Chapter 4

1 Vince Lombardi, "What it Takes to Be Number One," Vince Lombardi official website, accessed November 1, 2021, http://www.vincelombardi.com/number-one.html.

2 "habit," Merriam-Webster, accessed November 1, 2021, https://www.merriam-webster.com/dictionary/habit.

3 Will Durant, *The Story of Philosophy: The Lives and Opinions of the World's Greatest Philosophers* (New York: Pocket Books, 1991).

4   Colin Powell with Joseph Persico, *My American Journey* (New York: Random House, 2010).

5   Benjamin Franklin, *Benjamin Franklin's Book of Virtues* (Carlisle, MA: Applewood Books, 2016).

6   Vince Lombardi, "Famous Quotes by Vince Lombardi," Vince Lombardi official website, accessed November 1, 2021, http://www.vincelombardi.com/quotes.html.

7   Stephen R. Covey, *The Seven Habits of Highly Effective People* (Miami: Mango Publishing, 2015).

8   Theodore Roosevelt, *Citizen in a Republic* (Glasgow, Scotland: Good Press, 2020).

9   Abraham J. Malherbe, ed. *The Cynic Epistles: A Study Edition* (Atlanta: Society of Biblical Literature, 1977).

10 Laozi, *Tao Te Ching*.

Chapter 5

1   Diogenses Laërtius, *The Lives and Opinions of Eminent Philosophers* (CreateSpace Independent Publishing Platform, 2017).

2   Malcolm Gladwell, *Outliers: The Story of Success* (New York: Little, Brown & Company, 2008).

3   Sky News, "Anthony Joshua to 'reign again' after Ruiz rematch," YouTube video, November 23, 2019, https://www.youtube.com/watch?v=l_iGQcumZOY.

4   "Usain Bolt – The Real Work Is Done Behind the Scenes," YouTube video, May 18, 2019, https://www.youtube.com/watch?v=EOfGdbeAUIg.

5   Mike Tyson, *Undisputed Truth: My Autobiography* (New York, NY: HarperSport, 2014).

6   Atwood H. Townsend, *Good Reading: A Guide to the World's Best Books* (Mentor, 1948).

Chapter 6

1   Lucía Alejandro, José F. Dade las Heras, Margarita Pérez Margarita, Juan Elvira Juan, Alfredo Álvarez Antonio J. Carvajal, and José L. Chicharro MD, "The Importance of Physical Fitness In the Performance of Adequate Cardiopulmonary Resuscitation," *Clinical Investigations in Critical Care* 115, no. 1 (January 1999), https://doi.org/10.1378/chest.115.1.158.

Chapter 7

1   Albert Paine, *Mark Twain's Notebook* (London: Hesperides Press, 2006).

2   Discovery UK, "The Science Behind Lie Detector Tests | How Do They Do It?" YouTube video, March 31, 2018, https://www.youtube.com/watch?v=HwpTPAzgWjE.

3   Joe Navarro with Marvin Karlins, *What Every BODY is Saying: An Ex-FBI Agent's Guide to Speed-Reading People* (London: Collins, 2008).

4   Jordan Peterson, *12 Rules for Life, an Antidote to Chaos* (London: Allen Lane, 2018).

Chapter 8

1   Marcus Aurelius, *Meditations* (New York: Penguin, 2006).

2   James 3:5–6, New International Version (NIV).

3   Ephesians 4:29, NIV.

4   Luke 6:45, NIV.

Chapter 11

1   Maya Angelou; Jeffrey M. Elliot, ed., *Conversations with Maya Angelou* (Jackson: University Press of Mississippi, 1989).

2   Evan Carmichael, "SUCCESS Has NOTHING to Do With LUCK! | Michael Jordan | Top 10 Rules," YouTube video, July 26, 2015, https://www.youtube.com/watch?v=NidqtkXq9Yg.

3   Booker T. Washington, *Up From Slavery: The Incredible Life Story of Booker T. Washington* (Prague: Madison & Adams Press, 2018).

4   Friedrich Nietzsche, *Twilight of the Idols* (Oxford: Oxford University Press, 2008).

5   Kayla Stoner, "Science Proves That What Doesn't Kill You Makes You Stronger," Northwestern Now, October 1, 2019, https://news.northwestern.edu/stories/2019/10/science-proves-that-what-I-kill-you-makes-you-stronger/.

6   Henry Ford and Samuel Crowther, *My Life and Work* (Project Gutenberg, 2005).

7   Richard Feloni and Ashley Lutz, "23 Incredibly Successful People Who Failed At First," Business Insider, March 7, 2014, https://www.businessinsider.com/successful-people-who-failed-at-first-2014-3?r=US&IR=T.

8   Ibid.

9   Ibid.

10 Ibid.

11 "'You've got to find what you love,' Jobs says," Stanford News, June 14, 2005, https://news.stanford.edu/2005/06/14/jobs-061505/.

Chapter 12

1   "Arnold Schwarzenegger – 1997 A&E Biography with Jack Perkins," YouTube video, September 26, 2019, https://www.youtube.com/watch?v=EypYuyBZnpc&t=24s.

2   Jackie Robinson, *I Never Had it Made: An Autobiography* (New York: Ecco Press, 2013).

3   Duff McDonald, *Last Man Standing, The Ascent of Jamie Dimon and JPMorgan Chase* (New York: Simon & Schuster, 2010).

4   Nick Norman, "Famous Mentors and Mentees III – The Privilege of Pressure," Ye! Community, April 17, 2019, https://social.yecommunity.com/news/390398.

5   John C. Maxwell, *The Leader's Greatest Return: Attracting, Developing, and Multiplying Leaders* (New York: Harper Collins Leadership, 2020).

6   Martin Luther King Jr, *The Autobiography of Martin Luther King Jr.* (London: Abacus, 2000).

7   Ibid.

8   Ibid.

9   Ibid.

10 Robinson, *I Never Had it Made.*

11 Danny Heitman, "What did Bobby Kennedy do when the going got rough? He read," the *Christian Science Monitor*, October 17, 2016, https://www.csmonitor.com/Books/chapter-and-verse/2016/1017/What-did-Bobby-Kennedy-do-when-the-going-got-rough-He-read.

Chapter 13

1   Aimee Groth, "You're The Average Of The Five People You Spend The Most Time With," July 24, 2012, https://www.businessinsider.com/jim-rohn-youre-the-average-of-the-five-people-you-spend-the-most-time-with-2012-7.

2   Darren Hardy, *The Compound Effect: Jumpstart Your Life, Your Income, Your Success* (New York: Carroll & Graff, 2020).

3   Gay MacLaren, *Morally We Roll Along* (Boston: Little, Brown & Company, 1938).

Chapter 14

1   Anne Frank, *The Diary of a Young Girl* (US Imports, 2006).

Chapter 15

1   Ford, *My Life and Work.*

2   Marcel Schwantes, "15 Kobe Bryant Quotes From His Legendary Career That Will Inspire You," *Inc.*, January 26, 2020, https://www.inc.com/marcel-schwantes/15-kobe-bryant-quotes-from-his-legendary-career-that-will-inspire-you.html.

3   "Confucius – The Analects," USC US-China Institute, December 13, 1901, https://china.usc.edu/confucius-analects-17.

Chapter 16

1   Durant, *The Story of Philosophy*.

2   Judy Battista, "Tom Brady's seventh Super Bowl win the crowning achievement of his career," NFL, February 8, 2021, https://www.nfl.com/news/tom-brady-s-seventh-super-bowl-win-the-crowning-achievement-of-his-nfl-career.

Chapter 17

1   Nathaniel Hawthorne, *The Scarlet Letter* (Ware, UK: Wordsworth Editions, 1992).

2   Brian Tracy, *Maximum Achievement: Strategies and Skills That Will Unlock Your Hidden Powers to Succeed* (New York: Simon & Schuster, 1995).

3   Chris Kyle with Jim DeFelice and Scott McEwen, *American Sniper: The Autobiography of the Most Lethal Sniper in U.S. History* (New York: William Morrow, 2015).

4   "First Inaugural Address of Franklin D. Roosevelt," Yale Law School Lillian Goldman Law Library, March 4, 1933, https://avalon.law.yale.edu/20th_century/froos1.asp.

5   David Goggins, *Can't Hurt Me: Master Your Mind and Defy the Odds* (Carson City, NV: Lioncrest Publishing, 2018).

6   "Joe Rogan Experience #1080 – David Goggins," YouTube video, February 19, 2018, https://www.youtube.com/watch?v=5tSTk1083VY.

Chapter 18

1   Goggins, *Can't Hurt Me*.

2   Ibid.

3   Henry David Thoreau, *Walden* (Seattle: AmazonClassics, 2017).

Chapter 19

[1] "self-discipline," Dictionary.com, accessed November 4, 2021, https:// www.dictionary.com/browse/self-discipline.

[2] "IF ONLY YOU HAD DISCIPLINE + VISION – (powerful revelation) by Dr Myles Munroe," YouTube video, March 16, 2018, https:// www.youtube.com/watch?v=QNNir8_WPLs.

[3] "Will Smith – I Believe Self Discipline Is The Definition Of Self Love," YouTube video, March 4, 2018, https://www.youtube.com/ watch?v=260ou4_9mYI.

[4] Jocko Willink and Leif Babin, *Extreme Ownership: How U.S. Navy Seals Lead and Win* (New York: St. Martin's Press, 2017).

[5] Ibid.

[6] Nelson Mandela, *Long Walk to Freedom* (London: Abacus, 2015).

[7] "Newton's Laws of Motion," NASA Glenn Research Center, accessed November 4, 2021, https://www1.grc.nasa.gov/beginners-guide-to -aeronautics/newtons-laws-of-motion/.

[8] Ronda Rousey, *My Fight Your Fight* (New York: Random House, 2015).

Chapter 21

[1] G. Dautovic, "Smartphone Market Share: Past, Present and Future," Fortunly, June 10, 2021, https://fortunly.com/articles/ smartphone-market-share#gref.

[2] S. O'Dea, "Global smartphone market share from 4th quarter 2009 to 2nd quarter 2021," Statista, August 11, 2021, https://www.statista.com/ statistics/271496/global-market-share-held-by-smartphone-vendors- since-4th-quarter-2009/.

[3] Wolfgang Wild, "The golden age of the video store," Considerable, May 31, 2019, https://www.considerable.com/entertainment/retronaut/ video-rental-store-boom/.

[4] Frank Olito, "Rise and Fall of Blockbuster," Business Insider, August 20, 2020, https://www.businessinsider.com/rise-and-fall-of-blockbuster# in-2002-blockbusters-other-big-competitor-redbox-launched-9.

5   Ibid.

6   Ibid.

7   Ibid.

8   Ibid.

9   "Netflix, Inc. (NFLX)," Yahoo! Finance, accessed November 4, 2021, https://finance.yahoo.com/quote/NFLX?p=NFLX.

Chapter 22

1   Ford, *My Life and Work.*

2   "USPS vs Fedex (sic.) vs UPS," YouTube video, May 21, 2021, https://www.youtube.com/watch?v=tnHnfHUxTPw.

Chapter 23

1   Sharon A. Clark, "The Impact of the Hippocratic Oath in 2018: The Conflict of the Ideal of the Physician, the Knowledgeable Humanitarian, Versus the Corporate Medical Allegiance to Financial Models Contributes to Burnout," *Cureus* 10, no. 7 (2018), https://www.doi.org/10.7759/cureus.3076/.

2   Peter Tyson, "The Hippocratic Oath Today," PBS, March 26, 2001, https://www.pbs.org/wgbh/nova/article/hippocratic-oath-today/.

Chapter 24

1   Martin Luther King, Jr., "Martin Luther King, Jr.'s 'The Drum Major Instinct' Sermon Turns 50," Beacon Broadside, February 4, 2018, https://www.beaconbroadside.com/broadside/2018/02/martin-luther-king-jrs-the-drum-major-instinct-sermon-turns-50.html.

2   Ibid.

3   Kyle with DE Felice, McEwen, *American Sniper.*

4   "Major Richard 'Dick' Winters Tribute," YouTube video, September 15, 2011, https://www.youtube.com/watch?v=9sgJC_vQAL4.

5   Simon Sinek, *Start with Why, How Great Leaders Inspire Everyone to Take Action* (New York: Penguin, 2011).

6   Willink and Babin, *Extreme Ownership*.

7   "Walt Disney (DIS)," *Forbes*, accessed November 4, 2021, https://www. forbes.com/companies/walt-disney/?sh=394942d65730.

8   Robert Iger, *The Ride of a Lifetime: Lessons Learned From 15 Years as CEO of the Walt Disney Company* (New York: Random House, 2019).

9   Ibid.

10 "Jamie Dimon – Chairman and CEO of JPMorgan Chase," Coffee with the Greats podcast, July 14, 2020, https://anchor.fm/blamo-media/episodes/ Jamie-Dimon---Chairman-and-CEO-of-JPMorgan-Chase-egn5as.

11 Ford, *My Life and Work*.

12 Carl Jung, *Modern Man in Search of a Soul* (Milton Park, UK: Routledge Classics, 2001).

13 "Jamie Dimon – Chairman and CEO of JPMorgan Chase."

Chapter 25

1   Martin Luther King, Jr., *The Autobiography of Martin Luther King, Jr.* (London: Abacus, 2000).

2   "courage," Lexico by Oxford English Dictionary, accessed November 4, 2021, https://www.lexico.com/definition/courage.

3   John C. Maxwell, *Leaders Distinguish Themselves During Tough Times: Lesson 15 from Leadership Gold* (Nashville: Thomas Nelson, 2012).

4   Winston S. Churchill, "Finest Hour 133, Winter 2006-07," International Churchill Society, April 4, 2015, https://winstonchurchill.org/ publications/finest-hour/finest-hour-133/excerpts-from-the-sixth-churchill-lecture-winston-churchill-leadership-in-times-of-crisis/.

Chapter 26

1   Julie Broad, *The New Brand You: Your Image Makes the Sale for You* (Los Angeles: Stick Horse Publishing, 2016).

2   Matthew 13:57, English Standard Version (ESV).

Chapter 27

1 Matt Weinberger, "This is why Steve Jobs got fired from Apple – and how we came back to save the company," Business Insider, July 31, 2017, https://www.businessinsider.com/steve-jobs-apple-fired-returned-2017-7?r=US&IR=T.

2 "Steve Jobs' 2005 Stanford Commencement Address," YouTube, March 7, 2008, https://www.youtube.com/watch?v=UF8uR6Z6KLc.

3 Jonathan Haidt, *The Happiness Hypothesis: Putting Ancient Wisdom to the Test of Modern Science* (New York: Arrow, 2007).

Chapter 28

1 Oliver Wendell Holmes, *The Professor at the Breakfast-Table: With the Story of Iris* (Ithaca, NY: Cornell University Library, 2008).

2 Genesis 11:5–7, NIV.

3 Deborah Blagg and Susan Young, "What Makes a Good Leader?" Harvard Business School, April 2, 2001, https://hbswk.hbs.edu/item/what-makes-a-good-leader.

4 Elizabeth Bernstein, "Worried About a Difficult Conversation? Here's Advice From a Hostage Negotiator," the *Wall Street Journal*, June 14, 2020, https://www.wsj.com/articles/worried-about-a-difficult-conversation-heres-advice-from-a-hostage-negotiator-11592139600.

Chapter 29

1 Haidt, *The Happiness Hypothesis*.

Chapter 30

1 Lisa Woods, "3 Scientific Studies That Prove the Power of Positive Thinking," Medium, September 22, 2019, https://medium.com/swlh/3-scientific-studies-that-prove-the-power-of-positive-thinking-616477838555.

2   Barbara L. Fredrickson, Michael A. Cohn, Kimberly A. Coffey, Jolynn Pek, and Sandra M. Finkel, "Open Hearts Build Lives: Positive Emotions, Induced Through Loving-Kindness Meditation, Build Consequential Personal Resources," *Journal of Personality and Social Psychology* 95, no. 5 (2015), https://www.doi.org/10.1037/a0013262.

3   Ibid.

4   James Clear, "The Science of Positive Thinking: How Positive Thoughts Build Your Skills, Boost Your Health, and Improve Your Work," *Huffington Post*, updated September 25, 2017, https://www.huffpost.com/entry/positive-thinking_b_3512202.

5   "The Power of Positive Thinking," John Hopkins Medicine, accessed November 5, 2021, https://www.hopkinsmedicine.org/health/wellness-and-prevention/the-power-of-positive-thinking.

6   Ibid.

7   Abigail Johnson Hess, "New study from Stanford University finds that positivity makes kids more successful," CNBC, February 5, 2018, https://www.cnbc.com/2018/02/05/stanford-university-study-positivity-makes-kids-smarter.html.

8   Jessica Stillman, "New Stanford Study: A Positive Attitude Literally Makes Your Brain Work Better," *Inc.*, February 5, 2018, https://www.inc.com/jessica-stillman/stanford-research-attitude-matters-as-much-as-iq-in-kids-success.html.

9   "Anthony Robbins – A Habit of Positive Thinking," YouTube video, September 30, 2016, https://www.youtube.com/watch?v=ZB6yxZ5w1j8.

10 Chad M. Burton and Laura A. King, "The health benefits of writing about intensely positive experiences," *Journal of Research in Personality* 38, no. 2 (2004), https://doi.org/10.1016/S0092-6566(03)00058-8.

Chapter 31

1 Leo Tolstoy, *What is Art?* (New York: Penguin Classics, 1998).

2 "pride," Dictionary.com, accessed November 8, 2021, https://www.dictionary.com/browse/pride.

3 Ibid.

Chapter 32

1 "Issues in Jewish Ethics: Speech and Lashon HaRah," Jewish Virtual Library, accessed November 8, 2021, https://www.jewishvirtuallibrary.org/speech-and-lashon-harah.

2 "The Power Of Speech," The Online Hadracha Center, accessed November 8, 2021, https://www.hadracha.org/en/vw.asp?id=54.

3 "Speech and Lashon Ha-Ra," Jew FAQ, accessed November 8, 2021, https://www.jewfaq.org/speech.htm.

4 James 3:5–6, NIV.

5 Huda, "Lessons from the Qur'an Regarding Gossip and Backbiting," Learn Religions, January 6, 2019, https://www.learnreligions.com/gossip-and-backbiting-2004267.

6 Sam Woolfe, "A Buddhist Perspective on Gossiping," Sam Woolfe, March 19, 2015, https://www.samwoolfe.com/2015/03/a-buddhist-perspective-on-gossiping.html.

7 "Confucius, The Analects - 12," USC US-China Institute, December 13, 1901, https://china.usc.edu/confucius-analects-12.

Chapter 33

1 *Star Trek Voyager*, S1E2 "Parallax," 1995, 46 min.

2 Peter G. Northouse, *Leadership: Theory and Practice* (Thousand Oaks, CA: SAGE Publications, 2012).

3 Simon Sinek, *Start With Why: How Great Leaders Inspire Everyone to Take Action* (London: Portfolio, 2011).

4 Ibid.

5 Ibid.

6   Ibid.

7   "Jamie Dimon – Chairman and CEO of JPMorgan Chase."

8   The Rescue editors, "Bailed out banks," CNN Money, Accessed November 26, 2021, https://money.cnn.com/news/specials/storysupplement/bankbailout/.

9   Marcel Schwantes, "How Can You Tell Someone Has True Leadership Skills? This Legendary Football Coach Nails It With 1 Brilliant Sentence," *Inc.*, February 20, 2018, https://www.inc.com/marcel-schwantes/how-can-you-tell-someone-has-true-leadership-skills-this-legendary-football-coach-nails-it-with-1-brilliant-sentence.html.

10   Willink and Babin, *Extreme Ownership*.

11   Brené Brown, *Rising Strong: How the Ability to Reset Transforms the Way We Live, Love, Parent, and Lead* (New York: Random House, 2017).

Chapter 34

1   "Water Buffalo," National Geographic, accessed November 8, 2021, https://www.nationalgeographic.com/animals/mammals/facts/water-buffalo.

2   Ibid.

3   "Lion," Smithsonian's National Zoo & Conservation Biology Institute, accessed November 8, 2021, https://nationalzoo.si.edu/animals/lion.

4   Ibid.

Chapter 35

1   Frank, *The Diary of a Young Girl*.

2   Chief Justice John Roberts, "Cardigan Mountain School Commencement Speech," *Time,* June 3, 2017, https://time.com/4845150/chief-justice-john-roberts-commencement-speech-transcript/.

3   Viktor Frankl, *Man's Search for Meaning* (Boston: Beacon Press, 2006).

4   Ibid.

5   Ibid.

⁶ Friedrich Nietzsche, *Twilight of the Idols* (London: Penguin Classics, 1990).

Chapter 36

¹ Sinek, *Start with Why.*

² Ibid.

3 David Goldman, "Microsoft CEO Steve Ballmer to Retire," August 23, 2013, https://money.cnn.com/2013/08/23/technology/enterprise/microsoft-ballmer-retire/

⁴ Willink and Babin, *Extreme Ownership.*

Chapter 37

¹ Aleksandr Solzhenitsyn, Thomas P. Whitney, trans., *The First Circle* (London: Collins, 1968).

² Mark Miller, *Leaders Made Here: Building a Leadership Culture* (San Francisco: Berrett-Koehler, 2017).

³ John Maxwell, *The Leader's Greatest Return: Attracting, Developing, and Multiplying Leaders* (Nashville: HarperCollins Leadership, 2020).

⁴ Brown, *Dare to Lead.*

Chapter 38

¹ NBC Sports, "Tom Brady on being labeled the GOAT, Aaron Rodgers" YouTube video, November 3, 2018, https://www.youtube.com/watch?v=o8CrKP_3YPw.

Chapter 39

¹ Max Freedman, "How to Avoid Becoming a Micromanager," Business News Daily, September 4, 2020, https://www.businessnewsdaily.com/15794-avoid-becoming-a-micromanager.html.

Chapter 41

[1] Willink and Babin, *Extreme Ownership*.

Chapter 42

[1] Joel Stein, "Millennials: the Me Me Me Generation," *Time*, May 20, 2013, https://time.com/247/millennials-the-me-me-me-generation/.

[2] Brian Tracy, Dr. Peter Chee, *12 Disciplines of Leadership Excellence: How leaders Achieve Sustainable High Performance* (New York: McGraw-Hill Education, 2013).

[3] NBC Sports, "Tom Brady."

Chapter 43

[1] "Post-Traumatic Stress Disorder," *Psychology Today*, accessed November 10, 2020, https://www.psychologytoday.com/us/conditions/post-traumatic-stress-disorder.

[2] Matthew Tull, PhD, "Development of PTSD in Firefighters," Very Well Mind, February 7, 2020, https://www.verywellmind.com/rates-of-ptsd-in-firefighters-2797428.

[3] Ibid.

[4] Ibid.

[5] Ibid.

[6] Matthew Tull, "Coping with PTSD," Very Well Mind, June 30, 2021, https://www.verywellmind.com/coping-with-ptsd-2797536.

[7] James W. Pennebaker and Joshua M. Smyth, *Open Up by Writing it Down* (New York: The Guilford Press, 2016).

[8] Ibid.

[9] Seth J. Gillihan, "The Healing Power of Telling Your Trauma Story," *Psychology Today*, March 6, 2019, https://www.psychologytoday.com/us/blog/think-act-be/201903/the-healing-power-telling-your-trauma-story.

10 Bessel Van Der Kolk, *The Body Keeps the Score :Brain, Mind, and Body in the Healing of Trauma* (London: Penguin Books, 2015).

11 Ibid.

Commencement

1 Tim Fargo, "Who You Are Tomorrow is Decided Today | Social Jukebox," YouTube video, September 3, 2018, https://www.youtube.com/watch?v =M_uSbZiCJgI.

2 Tim Fargo, "Excuses Are The Rocks Where Our Dreams Are Crush," YouTube video, June 19, 2019, https://www.youtube.com/watch?v =6awyiKzoyjs.

3 Martin Luther King Jr., "Nobel Peace Prize Acceptance Speech," the Nobel Peace Prize, December 10, 1964, https://www.nobelprize.org/ prizes/peace/1964/king/26142-martin-luther-king-jr-acceptance-speech-1964/.

Appendix C

1 Johann Wolfgang Von Goethe; Joan DeRis Allen, trans., The Green Snake and the Beautiful Lily (New York: Rudolf Steiner, 2006).

2 1 Corinthians 13:4–8 NIV.

3 Viktor E. Frankl, Man's Search for Meaning (New York: Beacon Press, 2006).

ACKNOWLEDGMENTS

First and foremost, I want to thank God for his love and mercy and all the blessings in my life. I am eternally grateful to my loving wife and 3 kids for their love and their patience with me as I have written this book as well as for their thoughts, ideas and input. I could not have done it without you. You are the loves of my life, I am proud to be your husband and father and I hope to always make you proud of me.

I am also very grateful to the team that worked with me from Book Launchers. They helped take an extremely long and not well written document, by whittling and chipping away and adding a whole lot of polish and shine, to create the book I have today. I have learned so much along the way and I look forward to working with them in the future.

I want to thank all my instructors in the Fire Academy, EMT school, Paramedic school and the hundreds of other courses I have taken through the years for their diligent efforts to help train me and mold me into the firefighter I became. I am grateful for the late Vincent Elmore for taking a chance on a 21-year-old kid and bringing me into the most incredible occupation that exists, the Fire Service. I am also grateful that the residents of the Town of Palm Beach gave me an opportunity to serve them.

To those I worked alongside with, whom I grew to have deep friendships with, I say thank you very much, it was a blast working together, the memories I have, warm my soul and bring me joy whenever I think about them, life is nothing like "the lower east side." Thank you for watching my back and working hard with me to do an exceptional job and your continual friendship till this day. Most importantly, thank you for your love.

I extend my gratitude to Tony Robbins, Les Brown, and many others. There have been many large hurdles that have come up in my life. Your words motivated me to look inside of myself, to muster up courage and climb over those hurdles. I hope for your continued success with others, helping them overcome life's difficulties as well.

To all the men and women whose lives I read about, learned from and quoted and shared their example in this book, you have made such a significant difference in my life and the lives of my loved ones and I hope as well, the lives of those that will read and learn from this book. Thank you too for the many good things you have brought to the human family from which I too benefit from and the fine example you left for me to continue to follow. Thank you for your love of others.

www.ingramcontent.com/pod-product-compliance
Lightning Source LLC
Chambersburg PA
CBHW062114020426
42335CB00013B/959